AF356612

Stability and Seakeeping of Marine Vessels

Stability and Seakeeping of Marine Vessels

Editors

Ermina Begovic
Simone Mancini

MDPI • Basel • Beijing • Wuhan • Barcelona • Belgrade • Manchester • Tokyo • Cluj • Tianjin

Editors
Ermina Begovic
Department of Industrial
Engineering University of
Naples "Federico II"
Italy

Simone Mancini
Hydro and Aerodynamics
Department FORCE Technology
Denmark

Editorial Office
MDPI
St. Alban-Anlage 66
4052 Basel, Switzerland

This is a reprint of articles from the Special Issue published online in the open access journal *Journal of Marine Science and Engineering* (ISSN 2077-1312) (available at: https://www.mdpi.com/journal/jmse/special_issues/bz_stability_seakeeping_marine_vessels).

For citation purposes, cite each article independently as indicated on the article page online and as indicated below:

LastName, A.A.; LastName, B.B.; LastName, C.C. Article Title. *Journal Name* **Year**, *Volume Number*, Page Range.

ISBN 978-3-0365-0970-9 (Hbk)
ISBN 978-3-0365-0971-6 (PDF)

Cover image courtesy of Cameron Venti

Contents

About the Editors

Ermina Begovic graduated from Zagreb University, where she also finished her MSc degree. She completed her PhD at University of Naples Federico II, where, at present, she holds the position of Associate Professor of Ship Seakeeping. Her main research interests are high speed and planing craft seakeeping, safety and comfort on board of high speed ships, experimental hydrodynamics, hydrodynamic loads acting on intact and damaged ships, and high speed hull form design and optimization. Ermina Begovic is the author of more than 90 papers published in international journals, conference proceedings, and book chapters, and a member of the ISSC Specialist Committee on Special Craft and of Stability Research and Development Committee.

Simone Mancini, since 2020, has been the numerical team leader at the Department of Hydro and Aerodynamics at the FORCE Technology, Denmark. Furthermore, he was, for 15 years, a technical officer of the Italian Navy and, after periods of experience on board, he served as a project manager at the Ship Design Office of the Italian Navy General Staff. In 2012, he completed his MSc in Naval Architect and Marine Engineering at the University of Naples "Federico II", where he also completed his PhD in Computational Fluid Dynamics, in 2016. He is the author of more than 40 papers published in international journals, conference proceedings, and book chapters. He has served as a reviewer for more than 30 journals, and he is a member of the editorial board of two international journals. Since 2015, he has been a Contract Professor at the University "Giustino Fortunato". He is a SNAME member.

Preface to "Stability and Seakeeping of Marine Vessels"

Stability has always been the first safety issue for any marine vessel and static stability evaluation has been adequate for ship service. Recently, research interests have focused on ship dynamics and stability failure modes in rough sea for higher safety. Seakeeping assessment is today one of the most important design features to establish operational limits for both speed and comfort since the early design stage, thanks to the increasing of numerical and simulation capabilities. The new requirements on safety and comfort regulations make this analysis even more relevant. This book is based on the research papers of many scientists and engineers, showing the recent developments, especially in the application of numerical techniques in hull motions and added resistance evaluation and second generation intact stability criteria assessment.

Ermina Begovic, Simone Mancini
Editors

Journal of
Marine Science and Engineering

MDPI

Editorial

Stability and Seakeeping of Marine Vessels

Ermina Begovic [1],* and Simone Mancini [2]

[1] Department of Industrial Engineering, Università Degli Studi di Napoli Federico II, 80138 Naples, Italy
[2] Department of Hydro and Aerodynamics, Force Technology, 2800 Kgs. Lyngby, Denmark; simo@force.dk
* Correspondence: begovic@unina.it

Citation: Begovic, E.; Mancini, S. Stability and Seakeeping of Marine Vessels. *J. Mar. Sci. Eng.* **2021**, 9, 222. https://doi.org/10.3390/jmse9020222

Received: 16 February 2021
Accepted: 16 February 2021
Published: 19 February 2021

Publisher's Note: MDPI stays neutral with regard to jurisdictional claims in published maps and institutional affiliations.

Stability has always been the main safety issue for all marine vessels, and static stability evaluation is adequate for ship service. Recently, research interests have focused on ship dynamics and stability failure modes in rough seas for higher safety. Hence, seakeeping assessment today is one of the most important design features to establish operational limits for both speed and comfort.

The book *Stability and Seakeeping of Marine Vessels* includes nine contributions [1–9] to this Special Issue published during 2020. The overall aim of the collection is to improve knowledge about the most relevant and recent topics in ship stability and seakeeping. Specifically, the articles cover a wide range of topics regarding ship stability and seakeeping and reflect the recent scientific efforts in second-generation intact stability criteria evaluation and modelling of ship dynamics assessment in intact or damaged conditions. These topics are investigated mainly through direct assessments performed both via numerical methods and tools, such as boundary element methods (BEM) and computational fluid dynamics (CFD), and via EFD (experimental fluid dynamics). A brief overview of all the contributions, emphasizing the main investigation topic and the outcomes of the analysis, follows.

Zhang et al. (2019) [1] focus on the analysis of the effect of damage openings, specifically side and bottom damage openings, on the ship flooding process. The investigation is carried out numerically using the commercial CFD code CD Adapco Star CCM+, and the damaged hull investigated is the DTMB 5415 ship at zero speed. The URANS (Unsteady Reynolds average Navier–Stokes) simulation results indicate that the flooding process and the hull motion responses are strongly sensitive to the damaged opening positions and the internal compartment arrangement. Furthermore, the visualization of the flooding process efficiently explains the causes of the motion responses and can improve the analysis of ship survivability.

Mei et al. (2020) [2] propose an improved potential flow model for the hydrodynamic analysis of ships advancing in waves. This advanced potential flow method is based on the desingularized Rankine panel method, which is improved with the added effect of nonlinear steady wave-making (NSWM) flow in the frequency domain. The method is validated with results of simple geometry and a modified Wigley hull. A comparison of the results indicates that the improved model using the NSWM flow can give results in better agreement with the experimental results than those obtained using different potential flow approaches.

Pennino et al. (2020) [3] investigate a new adaptive weather routing model based on the Dijkstra shortest path algorithm with the aim of selecting the optimal route that maximizes the ship performances in a seaway. The model is based on a set of ship motion-limiting criteria and the weather forecast maps, providing the sea state conditions that the ship is expected to encounter along the scheduled route. The new adaptive weather routing model is applied to optimize the scheduled route of a containership in the northern Atlantic Ocean. The results show that it is possible to achieve appreciable improvements, up to 50% of the ship seakeeping performances, without excessively increasing the route length and the voyage duration.

Begovic et al. (2020) [4] present an experimental validation of the hybrid frequency–time domain method for vertical motions assessment for the hard-chine "low-drag" displacement hull. Currently, for a warped hard-chine hull operating in displacement and semi-displacement regimes, there is no adequate numerical tool available. Thorough validation of the applied method is performed in irregular head and following seas at Froude numbers Fr = 0.2, 0.4, and 0.6. Simulations are performed with the time step equal to the frequency of sampling, and an identical analysis of the numerical and experimental time series is performed, giving reliable results in head and following waves.

Petacco and Gualeni (2020) [5] give an overview of the development process that leads to the finalized version of second-generation intact stability criteria (SGIS), providing for each of the five stability failure modes a brief description of the vulnerability level requirements. Operational measures, categorized into two typologies—operational guidance and operational limitations—are analyzed through the application for a Ro-Ro Pax Ferry ship and finally provide guidance and limitations during navigation.

Ljulj and Slapnicar (2020) [6] present full-scale seakeeping tests conducted on a coastal patrol ship (CPS) during ship trials. Coast guards around the world have numerous challenges related to peacetime tasks, such as preventing human and drug tracking, fighting terrorism, controlling immigration, and protecting the marine environment, and, therefore, excellent seakeeping is of paramount importance. All relevant seakeeping criteria in different sea states are analyzed, including CPS behavior at high speed in severe sea states as well as the stern vertical displacement at the low speed required for launch and recovery tasks. The authors report a comparison of measured motions on sea against numerical calculations and model tests, and they review the set of seakeeping criteria to be used at the design stage.

Jing et al. (2020) [7] perform a study on the optimization of the motion response in waves of a barge platform using a zero-pressurized air cushion incorporated into the barge platform. The pressure of the zero-pressurized air cushion is equal to atmospheric pressure. Compared to the conventional pressurized air cushion, the zero-pressurized one has the advantage of less air leakage risk. The numerical results, based on the boundary element method, show that in regular and irregular waves, the air cushion could significantly reduce the amplitude of motions response close to the resonance condition.

Martic et al. (2020) [8] study the effect of ship characteristics on added resistance in regular waves and irregular head sea in relation to the IMO (International Maritime Organization) goal of reducing CO_2 emission by ships. Hydrodynamic calculations of added resistance of the KCS (Kriso container ship) are performed by the 3D panel method based on potential flow theory. The obtained numerical results are corrected for the direction component of added resistance in short waves and validated against the experimental data available in the literature. Numerical uncertainty is evaluated for the results in regular waves, whereas monotonic convergence is achieved and for the mean value of added resistance in irregular waves for certain sea states. The obtained results provide an overview of the effect ship characteristics variation on added resistance.

Finally, Pacuraru et al. (2020) [9] carry out a fully numerical analysis of the seakeeping performance of the KCS vessel. Several hydrodynamic methods, in-house code based on linear strip theory, the 3D fully nonlinear time-domain boundary element method (BEM), and the commercial CFD code NUMECA are employed to obtain accurate results of ship hydrodynamic response in regular head waves. The results obtained using these methods are presented and discussed to establish a methodology for estimating the ship response in regular waves with accurate results and to determine the sensitivity of hydrodynamical models.

Author Contributions: Conceptualization, E.B. and S.M.; resources, E.B. and S.M.; writing—original draft preparation, E.B. and S.M.; writing—review and editing, E.B. and S.M.; supervision, E.B. All authors have read and agreed to the published version of the manuscript.

Funding: This research received no external funding.

Conflicts of Interest: The authors declare no conflict of interest.

References

1. Zhang, X.; Lin, Z.; Mancini, S.; Li, P.; Liu, D.; Liu, F.; Pang, Z. Numerical Investigation into the Effect of Damage Openings on Ship Hydrodynamics by the Overset Mesh Technique. *J. Mar. Sci. Eng.* **2020**, *8*, 11. [CrossRef]
2. Mei, T.; Candries, M.; Lataire, E.; Zou, Z. Numerical Study on Hydrodynamics of Ships with Forward Speed Based on Nonlinear Steady Wave. *J. Mar. Sci. Eng.* **2020**, *8*, 106. [CrossRef]
3. Pennino, S.; Gaglione, S.; Innac, A.; Piscopo, V.; Scamardella, A. Development of a New Ship Adaptive Weather Routing Model Based on Seakeeping Analysis and Optimization. *J. Mar. Sci. Eng.* **2020**, *8*, 270. [CrossRef]
4. Begovic, E.; Bertorello, C.; Cakici, F.; Kahramanoglu, E.; Rinauro, B. Vertical Motions Prediction in Irregular Waves Using a Time Domain Approach for Hard Chine Displacement Hull. *J. Mar. Sci. Eng.* **2020**, *8*, 337. [CrossRef]
5. Petacco, N.; Gualeni, P. IMO Second Generation Intact Stability Criteria: General Overview and Focus on Operational Measures. *J. Mar. Sci. Eng.* **2020**, *8*, 494. [CrossRef]
6. Ljulj, A.; Slapničar, V. Seakeeping Performance of a New Coastal Patrol Ship for the Croatian Navy. *J. Mar. Sci. Eng.* **2020**, *8*, 518. [CrossRef]
7. Jing, F.; Xu, L.; Guo, Z.; Liu, H. A Theoretical Study on the Hydrodynamics of a Zero-Pressurized Air-Cushion-Assisted Barge Platform. *J. Mar. Sci. Eng.* **2020**, *8*, 664. [CrossRef]
8. Martić, I.; Degiuli, N.; Farkas, A.; Gospić, I. Evaluation of the Effect of Container Ship Characteristics on Added Resistance in Waves. *J. Mar. Sci. Eng.* **2020**, *8*, 696. [CrossRef]
9. Pacuraru, F.; Domnisoru, L.; Pacuraru, S. On the Comparative Seakeeping Analysis of the Full Scale KCS by Several Hydrodynamic Approaches. *J. Mar. Sci. Eng.* **2020**, *8*, 962. [CrossRef]

Journal of
Marine Science and Engineering

Article

Numerical Investigation into the Effect of Damage Openings on Ship Hydrodynamics by the Overset Mesh Technique

Xinlong Zhang [1], Zhuang Lin [1,*], Simone Mancini [2], Ping Li [1], Dengke Liu [1], Fei Liu [1] and Zhanwei Pang [1]

[1]   College of Shipbuilding Engineering, Harbin Engineering University, Harbin 150001, China;
hrbzhangxinlong@163.com (X.Z.); lp1355@163.com (P.L.); zhengdaliudengke@163.com (D.L.);
heuliufei@hrbeu.edu.cn (F.L.); zhanwei_pang@163.com (Z.P.)
[2]   Department of Industrial Engineering, University of Naples "Federico II", 80125 Naples, Italy;
simone.mancini@unina.it
*   Correspondence: linzhuang@hrbeu.edu.cn

Received: 8 November 2019; Accepted: 19 December 2019; Published: 23 December 2019

Abstract: Damage stability is difficult to assess due to the complex hydrodynamic phenomena regarding interactions between fluid and structures. Therefore, a detailed analysis of the flooding progression and motion responses is important for improving ship safety. In this paper, numerical simulations are performed on the damaged DTMB 5415 ship at zero speed. All calculation are carried out using CD Adapco Star CCM + software, investigating the effect of damage openings on ship hydrodynamics, including the side damage and the bottom damage. The computational domain is modelled by the overset mesh and solved using the unsteady Reynold-average Navier-Stokes (URANS) solver. An implicit solver is used to find the field of all hydrodynamics unknown quantities, in conjunction with an iterative solver to solve each time step. The Volume of Fluid (VOF) method is applied to visualize the flooding process and capture the complex hydrodynamics behaviors. The simulation results indicated that two damage locations produce the characteristic flooding processes, and the motion responses corresponding to the hydrodynamic behaviors are different. Through comparative analysis, due to the difference between the horizontal impact on the longitudinal bulkhead and the vertical impact on the bottom plate, the bottom damage scenario always has a larger heel angle than the side damage scenario in the same period. However, the pitch motions are basically consistent. Generally, the visualization of the flooding process is efficient to explain the causes of the motion responses. Also, when the damage occurs, regardless of the bottom damage or the side damage, the excessive heel angle due to asymmetric flooding is often a threat to ship survivability with respect to the pitch angle.

Keywords: URANS; VOF; overset mesh; side damage; bottom damage; flooding process; motion response

---

1. Introduction

Nowadays, ship safety is of high priority to the maritime industry. However, despite many efforts being to improve ship design in recent years, damage accidents continue to occur due to collision, grounding, or the unpredictable sea environment (wind, current and waves). The loss of hull integrity leading to damage flooding can be a severe risk to ship stability [1], even making the damaged ship sink or capsize. For a damaged ship, different damage scenarios correspond to the special opening locations. The resulting flooding processes and ship motion responses are also characteristic. In the flooding process, the ship motions impact the water flooding and sloshing in the flooded compartments. Simultaneously, the liquid loads acting on the compartments also influence the ship motions [2]. Therefore, the accurate prediction of the hydrodynamic behavior coupled with the ship motions is crucial to assess the remaining survivability of the damaged ship. Also, the complex hydrodynamic behaviors caused by the coupled motion has attracted significant attention at several recent International Towing Tank Conferences [3,4].

In order to enhance the understanding of the flooding process and motion responses of the damaged ship, a series of model experiments were performed while various numerical simulation methods were developed and implemented. An experimental campaign was carried out on a passenger ferry hull to underline the effects of the damage opening on the ship roll response. The damaged ship was placed in still water and beam regular waves at zero speed [5]. Lim et al. [6] used a course-keeping model ship to measure the advance speed and motion response of the damaged ship in head and following seas. Through the free-running tests, the motion characteristics under safe return to port (SRTP) regulations were identified. Siddiqui et al. [7] performed a detailed series of experiments in a wave flume on a thin walled prismatic hull form. The obtained results demonstrate the occurrence of sloshing and piston mode resonance in the tests and their influence on the hydrodynamics load of a damaged ship. The effect of air compressibility in the airtight compartment on local floodwater behavior was also investigated. Rodrigues, et al. [8] presented an experimental procedure to measure progressive flooding on a small-scale damaged floating body. Their focus was on the estimation of the discharge coefficient at different opening geometries in still water. Additionally, a numerical simulation based on the Reynolds averaged Navier-Stokes (RANS) solver was applied to validate its capability to reasonably reproduce the physical experiments. Korkut et al. [9] carried out six degrees of freedom motion responses tests in regular waves for intact and damaged conditions. The effect of the damage opening and waves with different wave heights and wave frequencies on the motion responses of the damaged model was explored. The obtained experimental results indicated that the damage opening has an adverse influence depending on the directionality of the waves and the applied wave frequency.

Generally, the applied experimental models above are created on the simplified assumption that the damaged compartment is empty, not considering the effect of permeability on the flooding and motion responses. In the real compartment layout, obstacles in the compartment and inner subdivision may affect the flooding path and quantity. Therefore, accounting for more realistic modeling of the damage flooding, Acanfora et al. [10] carried out an experimental investigation on the dynamic response of a damaged ship with a realistic arrangement of the flooded compartment. The results presented the effects of obstacles in the engine room compartment, such as decks and engine, on the roll responses. Similarly, Domesh et al. [11] used a damaged segmented ship model to study the effect of permeability and the internal arrangement of the damaged compartment on the pitch and heave responses. When the effect of the internal structures on the flooding process and motion responses is taken into account, the bearing capacity of different components will determine whether the secondary water ingress will occur. When the flooding water pressure exceeds the bearing limitation of components, the components will leak or collapse. Based on this aspect, Risto et al. [12] conducted unique full-scale tests to determine the leakage and collapse characteristics of various typical non-watertight structures, when subjected to the flooding water pressure. The obtained results can provide guideline values to determine when the structure may collapse with the accumulation of the flooding water. These well-designed model tests can accurately assess the damaged stability with complicated physical phenomena, establishing a database for the motion responses of the damaged ship.

Although the experimental tests can well investigate damaged ships, their flexibility and economic efficiency are very limited. In most cases, full-scale experiments are impossible, and model experiments are associated with problems of the scale effect. In the last decade, owing to the development of high-performance computers, there has been an increasing interest in the application of computational fluid dynamics (CFD) to investigate the multi-phenomena hydrodynamic problem of the damaged ship. A Navier-Stokes (NS) solver with a free surface capturing technique, i.e., the volume of fluid method, was developed to numerically simulate water flooding into a damaged vessel. The proposed method can be used to predict the dynamic behavior of the flooding water and its impact forces on the flooded compartment [13]. Sadat-Hosseini et al. [14] performed unsteady Reynold averaged Navier-Stokes (URANS) simulations for zero-speed damaged passenger ships in calm water and waves. The flooding procedure and roll decay in calm water were studied, and the motions in regular beam waves for various wavelength were analyzed. Even though the simulation demands a larger computation costs, the predicted results coincide better with the experiment results than those reported for potential flow solver. Santos et al. [15] described a mathematical model in the time domain of the motions and flooding of ships in a seaway. Different factors affecting the survivability of the damaged ships were assessed. Ming et al. [16] applied the weakly compressible smoothed particle hydrodynamics (SPH) method to explore the influences of transversal waves on the dynamic flooding process of a damaged compartment. The simulation results indicate that when the waves slam against the damaged ship, the relative position between the damage opening and the free surface will be changed. Further, different wave directions will result in different flooding processes. This method has the advantage of dealing with large deformation problems of free surface flow and fluid–structure interaction. In addition, Manderbacka et al. [17] and Acanfora et al. [18] presented a non-linear time domain simulation method for damaged ships. The flooding water motion is based on the lumped mass method with a moving free surface, and the ship's transient response to an abrupt flooding is simulated. It has been proven that these numerical methods have been an alternative approach to study the damage flooding.

In this paper, the URANS method combining the overset mesh technique in conjunction with the 6 Degree-of-Freedom (DOF) solver is applied to investigate the effect of side damage and bottom damage on the flooding process and motion responses. The paper is organized as follows. The overset mesh methodology is depicted in Section 2, including the definition of the overset mesh and the interpolation options. Section 3 introduces the utilized 5415 scale model. Section 4 underlies the whole simulation process and the critical settings, including the creation of the simulation domain, the boundary conditions, the choice of mesh types and the relevant solver settings. The captured flooding process and measured motion responses are presented and discussed in Section 5. Finally, conclusions and future research directions are taken in Section 6.

2. Overset Mesh Methodology

2.1. Definition of the Overset Mesh

With the development of computational fluid dynamics, time-domain simulation approaches based on the finite volume method (FOM) have been constantly evolving. However, for unstructured grids, the mesh-partitioning stage can be challenging due to the memory limitations of the massively parallel architectures. In this case, in order to run a simulation with increasingly fine grids and increasingly complex physics modelling, the high-performance computing represents a crucial capability to solve this problem [19]. Moreover, among the different motion-mesh techniques, the overset mesh (also known as Chimera or overlapping grids) has been considered as an efficient way to accurately describe the rotation motion of the damaged ship. Taking the simulation case as an example, as illustrated in Figure 1, when the overset mesh is activated, two individual regions are created: a background region and an overset region surrounding the damaged ship. The motion specification can be assigned to the overset region, rather than the background region. Because when the overset mesh is not applied, the motion specification can only be assigned to the background region. The background region will rotate relative to the stationary hull under the influence of the flooding water, monitoring the roll and pitch motion of the damaged hull. This will make the free surface out of the original mesh refinement block, resulting in poor simulation accuracy. However, the overset mesh can avoid this problem well. When the motion specification is assigned to the overset region, the background region is always stationary. The limited overset region will follow the movement of the damaged ship. Even if the damaged ship has a large roll or pitch motion, good overset mesh quality will ensure the accuracy of the simulation results. In addition, two regions are meshed separately, and the overset interface is created between them. For implicitly coupling the background region and the overset region, the interface is set to the overset mesh boundary condition. In this case, as the damaged ship moves within the background region, the overset region will correspondingly change. Simultaneously, the data information between the regions is exchanged through the overlapping cells. As illustrated in Figure 2, once the overset mesh is performed, the hole-cutting process in STAR-CCM + automatically couples the overset region with the background region through the overset interface. Four types of cells from the hole-cutting process are created. Active cells (cyan and yellow): Discretizing governing equations are solved here. Passive cells (dark blue). Donor cells (green): These provide interpolation information to the mesh acceptor cells. Acceptor cells (red): The boundary cells that receive information from the donor cells [20].

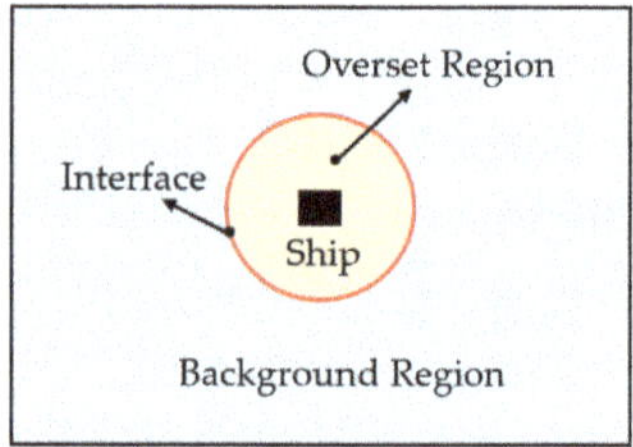

Figure 1. Sketch of the simulation domain.

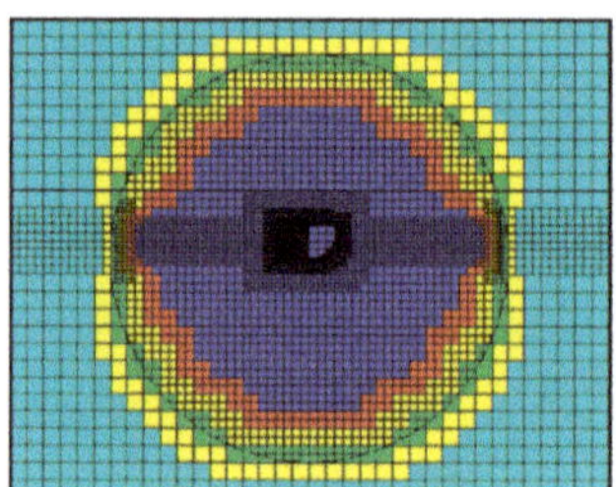

Figure 2. Four types of cells.

2.2. Interpolation Option

For the overset mesh, the interpolation function determines the data transfer relationship between the acceptor cells and the donor cells, ensuring implicit coupling of the background region and the overset region. In this case, a solution is computed on all grids simultaneously, leading to improved robustness and convergence. Of the specified interpolation options, the alternative interpolation options include distance-weighted interpolation, linear interpolation, and least-squares interpolation [21]. For the distance–weight interpolation, the interpolation factors are inversely proportional to the distance from the acceptor to the donor cell center, resulting in the closest cell giving the largest contribution. If the simulation involves moving mesh without great motion, the linear interpolation will be a better choice as it can ensure that interpolation elements do not overlap. This choice is more accurate but also more expensive due to the computational effort required. Finally, the least-squares interpolation transfers data from source to target meshes by data mappers. This method is suitable where there is a large variation of the moving grid with respect to the background mesh, as indicated in CD Adapco User's Guide [21] and De Luca et al. [22].

Based on the features of the simulation cases, linear interpolation is adopted to obtain an accurate solution.

3. Model Description

Numerical simulations have been performed using the well-known benchmark US Navy Destroyer Hull DTMB 5415 with a corresponding scale ratio of 1:25. Figures 3 and 4 respectively show the side view and body lines of the ship model, and Table 1 presents the principal dimensions of the ship model [23]. The created damaged compartment is located near the bow. The damaged compartment is assumed to be empty, not considering the influence of permeability and internal arrangements on the flooding water motion and damaged stability. However, in the real damage scenario, compartments could be full of equipment and obstacles that modify the flow of the water ingress. The simulation results will visualize the flooding process and measure the motion responses of the damaged ship in still water. The influence of the forward speed and external wave conditions on the coupled motion of the damaged ship and flooding water is not taken into account.

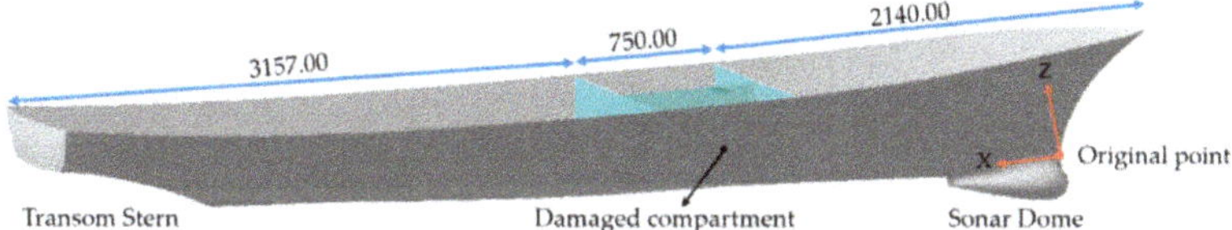

Figure 3. The schematic diagram of the damaged ship.

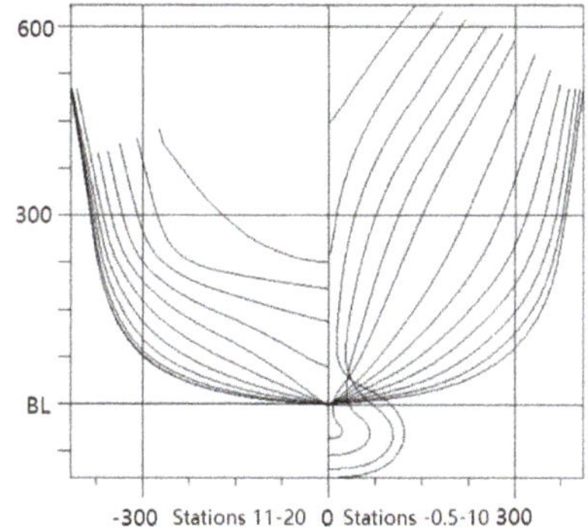

Figure 4. Body lines of DTMB 5415 model (Unit: inch).

Table 1. Principal dimensions of the DTMB 5415 model.

Parameters	Particulars	Real Ship	Scale Model (1/25)
Length Overall	L_{OA} (m)	151.1800	6.0470
Length between perpendiculars	L_{pp} (m)	142.0400	5.6856
Breadth at Waterline	B_{WL} (m)	20.0300	0.8012
Depth to public spaces deck	D (m)	12.7400	0.5096
Design draft	T (m)	6.3100	0.2524
Volume	V (m^3)	8811.94	0.5640
Maximum section area	A_X (m^2)	96.7923	0.1549
Block coefficient	C_B	0.4909	0.4909
Prismatic coefficient	C_P	0.6409	0.6409
Midship section coefficient	C_M	0.7658	0.7658
Height of metacenter above keel	KM (m)	9.4700	0.3788
Height of Centre of Gravity above keel	KG (m)	6.2830	0.2513
Metacentric height	GM (m)	3.1870	0.1272

As shown in Figure 5, two comparative damage scenarios are modeled separately. It can be found that the scenarios are specific to the location of the damage opening. Side damage and bottom damage will cause different types of flooding, while the corresponding motion responses are different. In the simulation process, in order to eliminate the composite influence arising from the air compressibility, an appropriate ventilation hole is constructed on the upper deck. In [24–27], it has been elaborated that the air compression in the flooded compartments will delay the flooding process and affect the dynamic behaviors of the damaged ship. According to the much-simplified assumption in MSC.362 (92) [28], if the total ventilation hole sectional area is 10% or more of the damage opening, the air compression may be neglected and the flooded compartment can be considered to be fully ventilated. Therefore, in order to ensure the adequate ventilation condition, a relatively large ventilation hole is set in the damage scenarios. The detailed dimensions of the damage opening and ventilation hole are shown in Table 2. It is worth noting that, except for the location of the damage opening, the other properties of the hull in the two scenarios are completely consistent, including the size and location of the ventilation hole, the size of the damage opening, the weight, the center of the gravity, and the inertia moments. In this case, it is meaningful to investigate the effect of the damage location on the flooding process and motion responses of the damaged ship. For the characteristics of the hull, a preliminary analysis was conducted by the means of a fine CAD (Computer aided design) model using CATIA CAD software. The density of the low carbon steel is assigned to each part in the CAD environment. Subsequently, the total weight of the hull, the center of mass, and moments of inertia can be accurately calculated, as presented in Table 2.

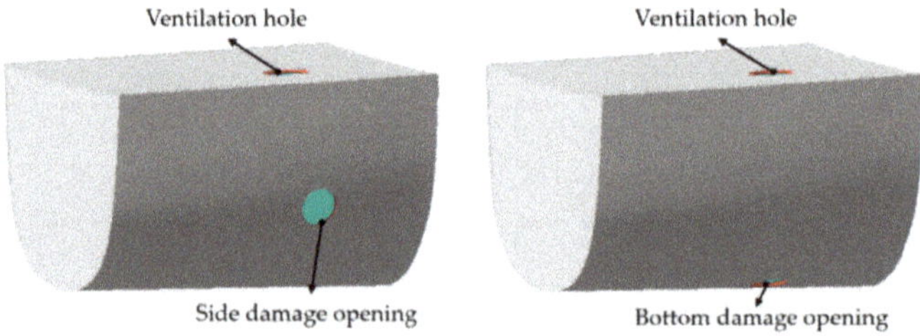

Figure 5. The schematic diagram of damage scenarios.

Table 2. Characteristics of the hull and details of the damage opening and the ventilation hole.

Parameters	Unit	Side/Bottom Damage Scenario	
Total Weight	**kg**	**665.64**	
Ventilation hole	mm	Radius	50
Damage opening	mm	Radius	40
Center of mass x	mm	2750.388	
Center of mass y	mm	−0.0260	
Center of mass z	mm	293.040	
Inertia moment I_{xx}	kg·m^2	60.3880	
Inertia moment I_{yy}	kg·m^2	1816.394	
Inertia moment I_{zz}	kg·m^2	1832.41	
Actual draft	m	0.269737	

After determining the characteristics of the damaged hull and ensuring that the damage location is the single variation, the actual draft of the damaged hull at the corresponding weight needs to be calculated. In the simulation setup, the height of the free surface needs to be consistent with the actual draft height calculated above. Specifying an accurate draft value will ensure that the damaged hull does not instantaneously heave due to the difference between weight and displacement at the moment of release, which also reproduces the actual physical process to some extent. Therefore, before carrying out the damage simulations, it is necessary to specify the weight of the damaged hull for the intact hull, calculating the actual draft in the case of the 665.64 kg displacement. As shown in Figure 6, the initial draft specified by the simulation is 0.3 m. Under the action of gravity and buoyancy, the final actual draft is stable at 0.269737 m. However, in the process of calculating the draft, only the vertical Z-axis motion is released, and other degrees of freedom are restrained. Such simplification will result in the transient pitch motion of the damaged hull due to the difference in bow and stern weight distribution. The specific simulation results will be analyzed in Section 5.

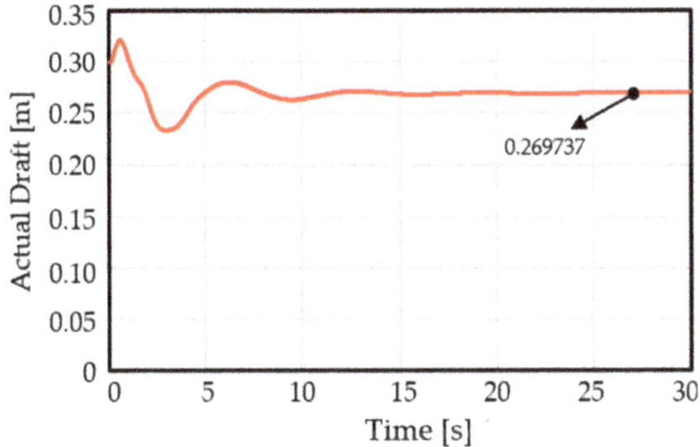

Figure 6. The actual initial draft of the damaged hull.

4. Numerical Setup

4.1. Simulation Domain and Physical Models

The applied overset mesh technique requires two different regions, including the background region and the overset region, as shown in Figure 7. The overset region is obtained by Boolean operation (subtraction) between the cylinder block and the damaged hull. The overset region will rotate and translate with the movements of the damaged ship. The background region is stationary, only providing the external flow field information. The background region and the overset region are implicit through the interface, while the connectivity between them takes place through the interpolation scheme specified for the interface. The interface mentioned here is the surface of the cylinder block. According to the Mancini et al. [29], Handschel et al. [30], and the available ITTC recommended procedure and guidelines [31], the dimensions of the background region and the overset region are summarized in Table 3. The background region is usually designed in compliance with the "Practical Guidelines for Ship CFD Application" [31]. However, there is no clear specification for the dimension of the overset region, as indicated by Tezdogan et al. [32]. It is worth mentioning that in the process of generating grids, the two regions use their individual mesh continuum to generate grids respectively, and the parameter properties of the two mesh continuums are independent of each other. However, in order to ensure the consistency of the external physical field, one identical physical continuum is applied to define the physical models in two regions.

Table 3. Presentation of domain dimensions.

Description	Symbol	Dimension	Mancini et al. [29]	Handschel et al. [30]
Domain length	a	$4.0L_{OA}$	$4.7L_{OA}$	$3.6L_{OA}$
Domain height	b	$3.0L_{OA}$	$2.7L_{OA}$	$1.8L_{OA}$
Domain breath	c	$3.0L_{OA}$	$3.4L_{OA}$	$1.2L_{OA}$
Inlet/outlet to cylinder	d,e	$1.2L_{OA}$	$1.7L_{OA}$	$1.2L_{OA}$
Cylinder to ship	h,g	$0.3L_{OA}$	$0.3L_{OA}$	$0.1L_{OA}$
Cylinder diameter	f	$3.75B_{WL}$	$4.7B_{OA}$	$2.0B_{WL}$

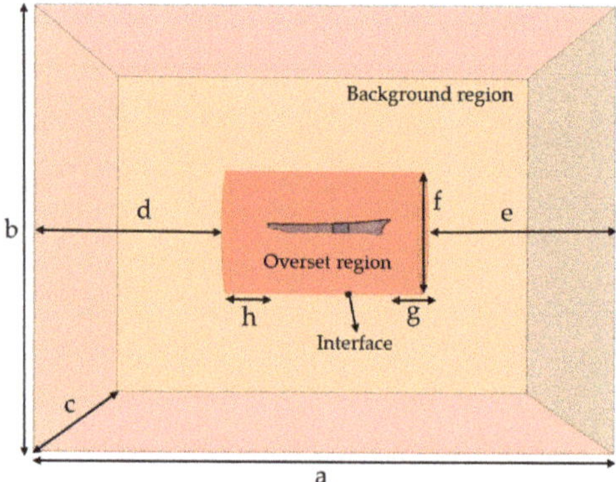

Figure 7. Region representations and domain dimensions.

In the simulations, the behaviors of two fluids (liquid and air) are modelled in the same physical continuum by the volume of fluid (VOF) approach. Due to the presence of two fluids, the Euler multiphase flow model is activated, and the gravity model is used to consider the gravitational effects of two fluids. The liquid phase is modeled with constant density water. In order to reproduce the real physical process, although the air compression is not considered, the air phase is still characterized by an ideal gas model. The necessary user defined field function (UDFF) model is needed to distribute the water and air [33]. Finally, a realizable k-ε two-Layer turbulence model is applied to solve the Reynolds stress problem, which can provide a good compromise between robustness, computational cost, and accuracy [34].

4.2. Boundary Conditions and Solver Settings

According to Zhang et al. [33] and Begovic et al. [35], the boundaries of the simulation domain are represented in Figure 8. The chosen boundary conditions and optimal solver settings are presented in Table 4. For a clear description of the internal arrangement of the simulation domain, the boundaries on both sides are not shown. It can be found that by creating isospheric surfaces, the entire simulation domain is divided into two parts by the free surface, including the air part above and the water part below. In order to obtain the sharp interfaces between the air and water, the second-order convection term is recommended. In this case, the high-resolution interface capturing (HRIC) scheme is designed to mimic the convective transport of immiscible fluid components, forming a scheme that is suited for tracking sharp interfaces.

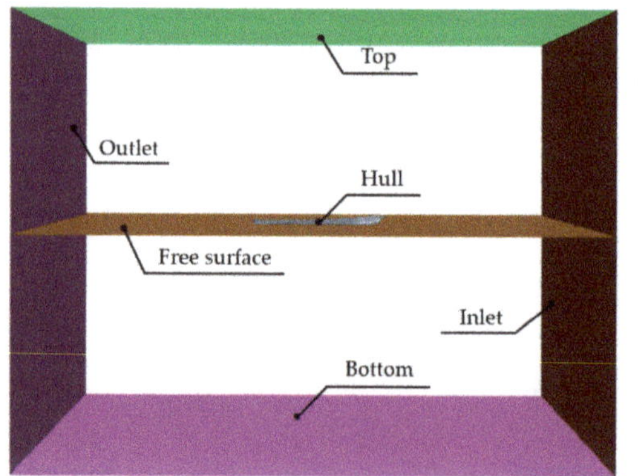

Figure 8. Boundaries representation of the simulation domain.

Table 4. Boundary conditions and solver settings.

Boundary Name	Boundary Type (This Paper)	Boundary Type Begovic et al. [35]	Boundary Type Zhang et al. [33]
Inlet	Velocity inlet	Velocity inlet	Velocity inlet
Outlet	Velocity inlet	Velocity inlet	Pressure outlet
Top/Bottom	Velocity inlet	Velocity inlet	Velocity inlet
Sides	Pressure outlet	Pressure outlet	Symmetry plane
Hull	Wall	Wall	Wall
Time step (s)	0.002	0.001	0.004
Maximum inner iterations	10	12	10
Convection Term	Second-order	Second-order	Second-order
Temporal Discretization	Second-order	Second-order	Second-order

The mass conservation equation and the momentum conservation equations including the turbulence model were calculated in the incompressible based unsteady state. For the coupling of velocity and pressure, a semi-implicit method for pressure linked equations (SIMPLE) method was used [36]. In order to increase the convergence performance of the linear algebraic equation, AMG (Algebraic Multi-Grid) method [37] was used and, using the Gauss-Seidel method, the simultaneous linear equation was solved.

The URANS (unsteady Reynolds-averaged Navier-Stokes) equations have been applied to control the update at each physical time for the calculation. In order to converge the solution for that given instant of time, each physical time is set to involve some number of inner iterations. Considering the compromise between the computational accuracy and the computational cost, the simulation program in this paper used a constant time step of 0.002 s, while the maximum number of the inner iteration steps is 10. The numbers determined are also consistent with the related recommendations of practical guidelines for ship CFD applications [31]. In addition, the second-order temporal discretization is activated to perform the transient calculations, which uses the current time level and the solutions from the previous two-time levels. Therefore, when the solver performs first-step calculation utilizing the second-order temporal discretization, the first-order temporal discretization is temporarily activated to provide the solutions of the first two inner iteration steps.

4.3. Mesh Type and Mesh Size

For the mesh type, the trimmed hexahedral type is used to generate the mesh. In Begovic et al. [35], detailed sensitivity analysis of the mesh types has been performed on the roll damping assessments of the damaged ship with two hybrid meshes (polyhedral and trimmed) and two trimmed meshes. The simulation results indicate that the hybrid meshes are prohibitive due to the high time consumption and poor simulation accuracy, while the trimmed meshes are recommended. Based on this conclusion, the generated mesh in this paper is shown in Figure 5. It can be found that the entire domain is divided into three regions with different mesh densities, including the background region, the cylindrical overset region, and the overlapping region. In order to optimize the discretization of the overset region, the mesh density of the overset region is denser than the other two regions. To minimize the errors that occur when interpolating variables between two meshes, the same mesh density order of magnitude is used in the overlapping region and the background region. Therefore, it can be seen from Figure 9 that the mesh density of the overlapping region is denser than the background region and coarser than the overset region. However, fundamentally, the overlapping region is still part of the background area. It is just a refinement block extracted from the background region. In addition, the meshes around the free surface are also locally refined, which can prevent the floating-point exception due to the free surface fluctuation or breakage. Finally, in order to avoid large computational costs, the free surface of the overset region is finer than that of the background region. The mesh sizes in different parts are summarized in Table 5.

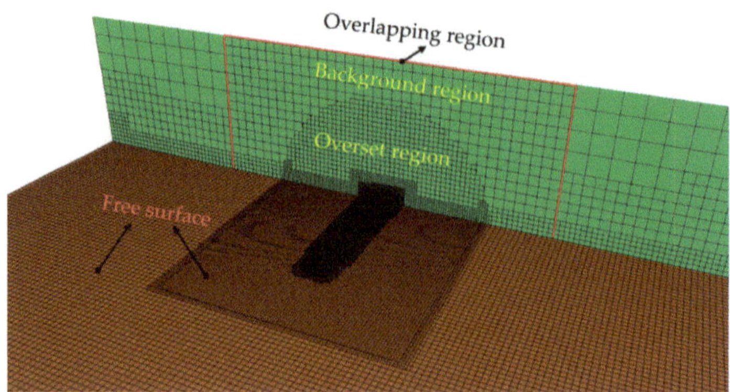

Figure 9. Visualization of the hexahedral mesh.

Table 5. Mesh sizes in different parts.

Part	Affiliated Region	Wrapper Size	Remesh Size	Trim Size
Bulbous bow	Overset region	0.050 m	0.015 m	0.110 m
Damage opening	Overset region	0.050 m	0.010 m	0.110 m
Ventilation hole	Overset region	0.050 m	0.010 m	0.110 m
Ship	Overset region	0.100 m	0.015 m	0.110 m
Flooded compartment	Overset region	0.060 m	0.015 m	0.010 m
Overlapping region	Overset region	0.100 m	0.250 m	0.110 m
Overlapping region	Background region	None	0.250 m	0.250 m
Free surface	Overset region	0.100 m	0.250 m	0.060 m
Free surface	Background region	None	0.250 m	0.120 m

In addition to optimizing the mesh sizes of the different regions described above, the mesh quality of the damaged ship also determines the calculation accuracy. As shown in Table 5, the wrapper model, the remesh model and the trim model are activated in the mesh continuum. The appropriate wrapper and remesh sizes will restore the original geometry of the damaged ship. It can also be found that the wrapper model is not activated in the mesh continuum of the background region. This is because the damaged ship is located in the overset region, and only the wrapper characteristics in the mesh continuum of the overset region can be defined to the damaged ship. The setting of a mesh size requires a comprehensive combination of mesh quality and the mesh number. Too fine meshes will cause a longer computation time, while too coarse meshes will result in poor computation convergence. Taking the mesh size of the overlapping region as an example, when the trim size is set to 0.100 m, the mesh number will be much larger than when the mesh size is set to 0.110 m. Consequently, the time consumption is very high. So, in order to achieve the balance between the simulation accuracy and the time consumption, the final trim size is to set 0.110 m. Similarly, the mesh sizes of other parts are determined after repeated attempts. Especially for the damage opening, the ventilation hole, and the bulbous bow, since their outer contours display a large curvature, the wrapper and remesh sizes need to be modified repeatedly to ensure the real opening shape. For the flooded compartment, the trim size is locally refined. Because only when the mesh is fine enough, the complex hydrodynamics phenomena in the flooding process can be accurately captured. The generated surface of the volume mesh is shown in Figure 10. It can be seen from the figure that the original surface of the damaged ship is restored well, and the necessary block is correspondingly refined.

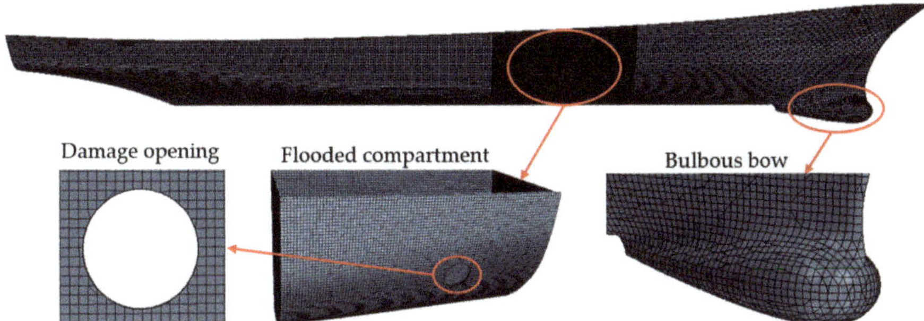

Figure 10. Volume mesh of the hull and refinement details of the local blocks.

4.4. Near-Wall Treatment

The wall function approach is used for the near-wall treatment, in particular, the All wall Y + model. This approach is formulated to assure reasonable answers for meshes of intermediate resolution and is considered as the best compromise between description of the boundary layer with acceptable quality and the time required for the calculation [29]. The wall y + is a non-dimensional distance similar to the local Reynolds number, often used in CFD to describe how coarse or fine a mesh is for a particular flow [38]. As indicated in the User's Guide [21], values of y+ $\approx$ 30 are most desirable for wall functions, whereas values of y+ $\approx$ 1 are most desirable for near-wall modeling. The values of wall y + on the hull surface is shown in Figure 11. It can be found that the y + values on the hull are very close to 1. For this reason, the realizable k-ε two layers turbulence model is applied. This turbulence model represents an improved treatment of the near-wall region for turbulent flows at low Reynolds numbers. This model is characterized for the layer next to the wall, where the turbulent dissipation rate and the turbulent viscosity are specified as functions of wall distance. More details about this model are available in Rodi [39].

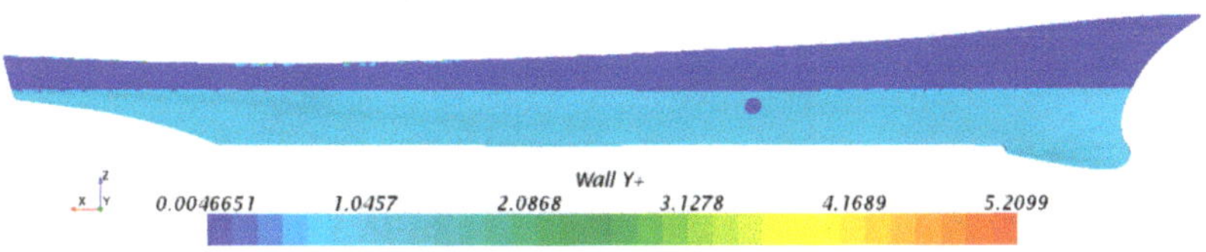

Figure 11. Wall Y + visualization on the hull.

5. Simulation Results and Discussion

Based on the appropriate solver settings and the optimized mesh generation, the simulation results respectively analyze the effect of damage locations on the flooding process and the motion responses. The complex hydrodynamic behaviors in the flooding process are visualized. The corresponding motion responses are compared and discussed, including the roll and pitch motion.

5.1. The Analysis of the Flooding Process in the Side Damage Scenario

Figure 12 presents the distribution of the flooding water at different time points. In the early stage of the damage flooding, due to large pressure difference between inside and outside of the damaged opening, the water ingress flows into the flooded compartment with a jet form in the opening section. The water ingress impacts the bottom plate and the longitudinal bulkhead, resulting in the splashing of the flooding water. Since the flooding water is violent, the compressed air in the damaged compartment cannot smoothly escape from the ventilation hole, and the complex hydrodynamic behavior such as a bubble is formed in the flooding water. This stage is complex but short, which is often defined as the transient flooding stage. As shown in Figure 12, the period from 0 s to 4.9 s can be roughly referred to as the transient flooding stage. With the effect of asymmetric water ingress, the damaged ship heels towards the starboard side, causing the pressure at the opening section to become larger. From the four graphs corresponding to 1, 2.9, 4, and 4.9 s, it can be found that the increased pressure at the opening section makes the slamming point (1, 2, 3, 4) on the longitude move upward. Such a slamming effect will produce a restoring moment that makes the damaged ship heel towards the port side. When the flooding water develops to a certain extent, the pressure difference between the inside and outside of the damaged compartment will gradually smaller, the flooding will become slower. This stage is often referred to as the progressive flooding stage, as shown in Figure 12 for the period from 4.9 s to 20 s. In this stage, the flooding water continuously flooded the damaged compartment. The free surface exhibits a wave-propagating form, producing a reflective behavior when it touches the longitudinal bulkhead. Although the flooding process is almost completed in about 20 s, from the capture of the flooding process at 30 s, the free surface is still sloshing due to the roll motion of the damaged ship. Conversely, the sloshing of the free surface also affects the motion response of the damaged ship. Finally, if the damaged ship can keep afloat, not capsizing or sinking due to the added flooding water, the final equilibrium state will be characterized. This stage is often referred to as the steady stage. Such detailed descriptions of the hydrodynamic behaviors in the flooding process can be applied to explain the causes of the damaged ship's motion responses. The specific and comprehensive explanation is elaborated in the motion response analysis in Section 5.3.1.

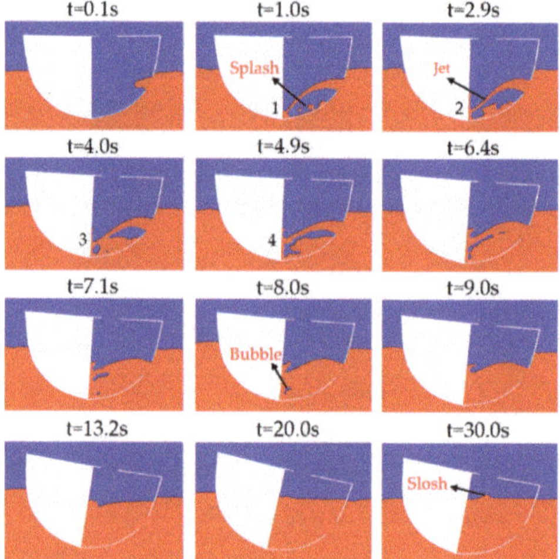

Figure 12. Visualization of the side flooding process.

5.2. The Analysis of the Flooding Process in the Bottom Damage Scenario

After comparison, there are both similarities and differences between the side flooding process illustrated in Figure 12 and the bottom flooding process illustrated in Figure 13. The overall similarity is that the bottom flooding process also experiences three flooding stages, including the transient stage, the progressing stage, and the steady stage. In the transient flooding stage (0–2.0 s), due to the large pressure difference between the inside and outside of the damaged opening, the violent seawater flooded the damaged compartment in a short time. Then, with the accumulation of the flooding water, the pressure difference gradually reduces while the flooding water occupied the damaged compartment at a slow rate (2.0–15.2 s). Finally, under the coupled influence of the tank sloshing, the damaged ship tends to be stable in the roll decay motion (15.2 s–30.0 s). At 24.6 s and 30.0 s, the internal wave propagation caused by the sloshing of the free surface can be clearly seen in the damaged compartment. At the same time, the flooding characteristics in the bottom damage scenario are also evident. Corresponding to the normal direction of the bottom opening section, the flooding water is sprayed from the ship bottom in the form of a water column. When the flooding water slams the longitudinal bulkhead and reaches the highest point at 0.3 s, the flooding waterfalls under the influence of gravity. Complex dynamic behaviors can be observed when the falling flooding water touches the ship bottom from 0.5 to 1.2 s, including splashing and bubble phenomena. In addition, in contrast to the side damage scenario, the damage opening in the bottom damage scenario is located deeper below the waterline. In this case, the hydrostatic pressure at the bottom opening section is much greater than that at the side bottom opening section. This also explains why under the premise of the same damage opening size, the bottom flooding process is completed in about 15 s, however, the side damage flooding takes about 20 s. Although the extra 5 s is relatively short, according to the Froude law, the converted flooding time for the real ship longer is than on the scale model. All simulation cases in this paper were carried out on the scale model (1/25). Therefore, the extra 5 s is about 25 s when converted to the full-scale ship. Once the damage accident occurs, 25 s can provide more rescue options. Therefore, accurately predicting the flooding time in different damage scenarios is meaningful for the emergency crew to take appropriate rescue managements. Generally, this detailed visualization of the hydrodynamic behaviors is helpful to enhance the understanding of the entire flooding process among the crew and ship designers.

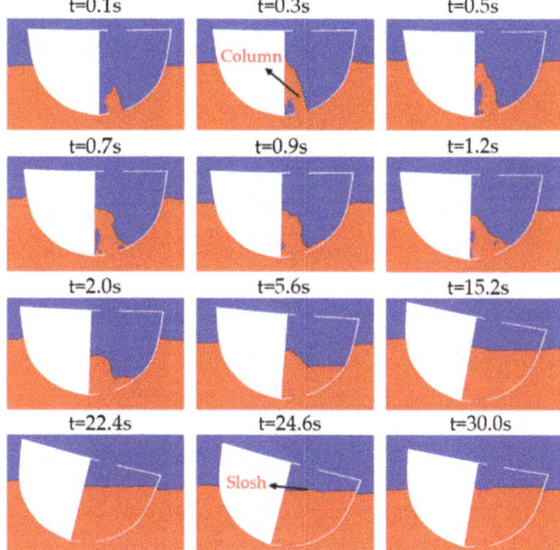

Figure 13. Visualization of the bottom flooding process.

5.3. The Analysis of the Coupled Motion Responses

5.3.1. Description of the Roll Motion Response

Based on the hydrodynamic behavior in Section 5.1. and Section 5.2, the resulting motion responses are elaborated in this section. As shown in Figure 14, although the damage locations of the two damage scenarios are different, the asymmetric flooding occurs in both damage scenarios. The asymmetric moment generated by the flooding water causes the damaged ship to heel only in the starboard. There is no periodic reciprocating roll motion between the portside and starboard. According to the right-hand rule, the value of the heel angle is negative. The heel angle can reach 15 degrees or more, indicating that the asymmetric flooding in a damaged ship is a dangerous situation. Therefore, efficient and feasible cross-flooding arrangements are needed to provide the necessary equalization across the ship in order to decrease the heel angle [40]. After a separate analysis of the roll motion curves in the two damage scenarios, it can be found that the damaged ship does not always heel towards the starboard, but gradually heels during left and right shaking. On the one hand, this is due to the inherent restoring moment of the hull itself. On the other hand, the flooding water causes a leftward impact on the longitudinal bulkhead, which also causes the damaged ship to have a tendency to heel towards the portside. As can be seen from Figure 12, the side flooding water continuously impacts the longitudinal bulkhead from the starboard. However, the horizontal impact effect due to side flooding is small compared to the vertical effect of the flooding water accumulation on the starboard bottom plate. In this case, the damaged ship only has a slight tendency to heel towards the portside, and never has a positive heel angle. At the same time, since the normal direction of the bottom opening section points to the longitudinal bulkhead, the water column in Figure 13 inevitably have an impact effect on the longitudinal bulkhead, so that the same shaking phenomenon as the side damage scenario occurs in the roll motion curve of the bottom damage scenario. In addition, the roll motion of the damaged ship and the flooding water affect each other. The roll motion makes the flooding water slosh in the flooded compartment. Conversely, the water sloshing has an impact effect on the internal bulkhead, including the longitudinal bulkhead and hull plate. Such coupled motion can present a risk to the ship survivability, even making the damaged ship capsize due to the parametric roll motion. This coupled analysis also explains why the damage scenarios still have a roll motion, even though the flooding water is no longer increased. However, due to the dissipation of energy, the entire roll motion will gradually decay.

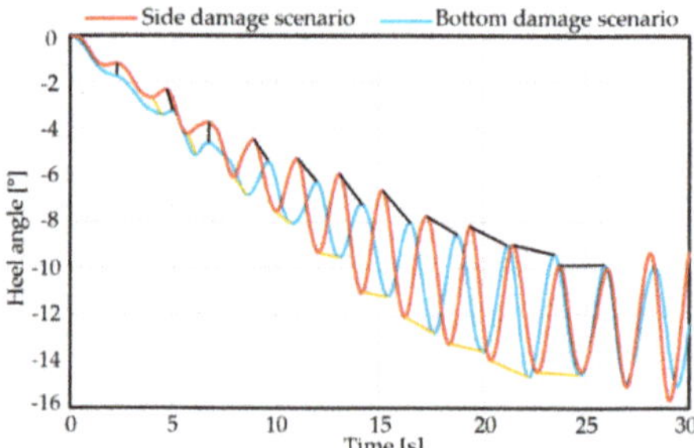

Figure 14. Comparison of heel angles in different damage scenarios.

Finally, by comprehensively comparing and analyzing the roll motion of the two damage scenarios, it can be found that different hydrodynamic behaviors produce distinct motion responses. The roll motion curves of the two damaged scenarios follow the similar periodic variation rule. However, there are certain differences in the peak and trough values for the same period. In Figure 14, the peaks and troughs in the same periods are connected by the black line segments and the yellow line segments. The magnitude of the peak value represents the extent to which the damaged ship heels towards the portside (the intact side), and the magnitude of the trough value represents the extent to which the damaged ship heels towards the starboard (the damaged side). It can be seen that the bottom damage flooding in the same period produces a larger heel angle with respect to the side damage flooding, regardless of the peak value or trough value. The reason for this difference is close to the hydrodynamic behaviors of the specified damage scenario. For the bottom damage flooding, the vertical effect of the upward flooding on the bottom plate is much larger than the horizontal effect of the water column on the longitudinal bulkhead. In this case, in comparison with the side damage scenario, the damaged ship with the bottom opening has a greater inclination towards the starboard. This also corresponds to the fact that the trough values of the bottom damage scenario are below those of the side damage scenario. For the side damage flooding, the flooding water strikes the longitudinal bulkhead from the starboard to the portside. The horizontal impact drives the damaged ship heel towards the portside. This explains why the peak values of the side damage scenario in the same period are above the bottom damage scenario. In general, the visualization of the flooding process is very helpful and meaningful for analyzing the causes of motion responses.

5.3.2. Description of the Pitch Motion Response

For analyzing the influence of the damage location on the pitch motion, Figure 15 presents the pitch motion curves of the two damage scenarios. From the overall analysis, the maximum values of the pitch angle for the two damage scenarios are only about 2 degrees in the flooding process. Conversely, the maximum value of the heel angle can reach about 15 degrees. This validates a basic conclusion that the damaged ships rarely lose stability due to the excessive pitch angle. The damaged ships often capsized due to the excessive heel angle caused by the additional flooding water. Therefore, this also provides an empirical reference for the ship designers and the emergency personnel onboard. When the damage occurs, especially for asymmetric flooding, in order to reduce the risk of capsizing or sinking, appropriate countermeasures should be taken to equalize the heel angle.

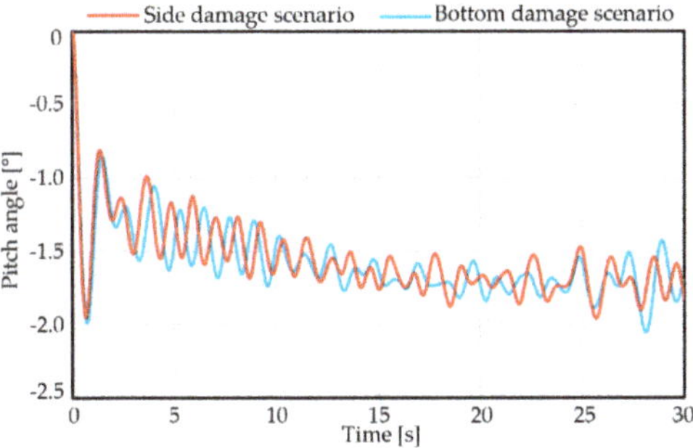

Figure 15. Comparison of pitch angles in different damage scenarios.

Although the pitch angle caused by the flooding water does not pose a threat to the safety of the damaged ship, its relevant effect deserves enough attention. Especially for the military ships, even if the ship is in a damaged situation, the ship must guarantee the corresponding operational missions and has to able to recover functionality following an incident (recoverability). Hence, the resulting 2 degree pitch angle is likely to affect the accuracy of the weapon strikes. So, the way to eliminate the extra pitch angle is of research significance. By comparing the pitch motion curves of the two damage scenarios, it can be found that the two pitch motion curves maintain the same variation rule, and the differences of the peak and trough values are small. This shows that the effect of the damage location on the pitch motion is small, even can be negligible. For analyzing the tendency of the pitch motion curves, it can be divided into two parts. The first part is that the damaged ship will have a transient head-pitching process at the beginning of the calculation. Because in the simulation settings, the damaged ship is placed horizontally by default. When the simulation runs, the damaged ship firstly have the initial pitch motion due to the uneven distribution of the fore and bow weights. Therefore, it can be seen from the Figure 15 that the pitch motion curves of the two damage scenarios are basically consistent at the beginning of the damage flooding, which is caused by the ship's own weight distribution described above and has little to do with the hydrodynamic behavior. The second part is that, as the flooding continues to develop, the flooding water gradually flooded the damaged compartment. Due to the strong nonlinear phenomenon of the flooding water, the flooding water spread irregularly in the damaged compartment. This is why the predicting pitch motion curves fluctuate up and down. However, the amplitude of the peak and trough values are very small. In this case, the crew onboard will not have the sloshing feeling caused by the pitch motion. Therefore, from the perspective of ensuring the survivability of the hull and the safety of human life, the threat posed by the roll motion is the problem that should be solved first.

6. Conclusion and Future Researches

The paper demonstrated the feasibility of CFD simulations to investigate complex flooding phenomena. The developed numerical approach in this paper can well capture the complex hydrodynamic behavior in the flooding process, including splash, jet, water column, and bubble. The URANS solver involving the overset mesh technique is applied to monitor the motion responses of the damaged ship with different damage locations. Through analysis, the visualization of the flooding process can be efficient to explain the cause of the resulting motion response. Comparing the bottom damage and the side damage, the upward flooding in the bottom damage scenario causes a larger vertical impact on the bottom plate while the flooding water from starboard to portside in the side damage scenario causes a larger horizontal impact on the longitudinal bulkhead. This detailed visualization description explains why the bottom damage flooding in the same period produces a larger heel angle with respect to the side damage flooding, regardless of the peak value or trough value. In addition, due to the coupled motion between the damaged ship and the flooding water, the tank sloshing makes the damaged ship still roll even if the flooding water is not increased. And, the wave propagation in the flooded is seen clearly. After comparing the roll motion and pitch motion, it can be found that the symmetric flooding is a dangerous situation, even making the damaged ship capsize. Though the damage locations are different, the asymmetric flooding produces an excessive heel angle. However, asymmetric flooding has little effect on the pitch angle. Based on these summaries, the final suggestion is that when the damage flooding takes place, especially asymmetric flooding, appropriate counter measures should be taken first to equalize the heel angel. In this case, the damaged ship can keep a good floating state, which is helpful to improve the survivability of the damaged ship and ensure the safety of human life. From the perspective of the simulation validation, only numerical simulation results are introduced, without validation with experimental results. Subsequently, the specified model tests will be carried out to prove the reliability of the applied numerical simulation approach.

In the future, a more realistic compartment arrangement will be created, considering the effect of the permeability on the flooding process and motion responses. Also, more sea conditions will be included, including wind, wave, and forward speed. This gradual improvement process will also present new challenges to the current simulation approach.

Author Contributions: X.Z. performed the numerical simulation and wrote the paper. Z.L. gave the investigating idea and revised the manuscript. S.M. optimized the numerical program and revised the manuscript. P.L. provided the detailed model information. D.L. established the damaged 5415 model. F.L., and Z.P. post-processed the simulation results. All authors have read and agreed to the published version of the manuscript.

Funding: This research was funded by the National Natural Science Foundation of China (NSFC Grants 51709063).

Acknowledgments: This research was supported by the College of Shipbuilding Engineering, Harbin Engineering University.

Conflicts of Interest: The authors declare no conflict of interest.

References

1. Manderbacka, T.; Themelis, N.; Bačkalov, I.; Boulougouris, E.; Eliopoulou, E.; Hashimoto, H.; Konovessis, D.; Leguen, J.F.; González, M.M.; Rodríguez, C.A.; et al. An overview of the current research on stability of ships and ocean vehicles: The STAB 2018 perspective. *Ocean. Eng.* **2019**, *186*, 106090. [CrossRef]
2. Gao, Z.; Gao, Q.; Vassalos, D. Numerical simulation of flooding of a damaged ship. *Ocean. Eng.* **2011**, *38*, 1649–1662. [CrossRef]
3. ITTC. The specialist committee on prediction of extreme ship motions and capsizing. In Proceedings of the 27th International Towing Tank Conference, Copenhagen, Denmark, 31 August–5 September 2014.
4. ITTC. Stability in Waves Committee, Final Report and Recommendation. In Proceedings of the 28th International Towing Tank Conference, Wuxi, China, 17–22 September 2017.
5. Acanfora, M.; De Luca, F. An experimental investigation into the influence of the damage openings on ship response. *Appl. Ocean. Res.* **2016**, *58*, 62–70. [CrossRef]

6. Lim, T.; Seo, J.; Park, S.T.; Rhee, S.H. Free-running Model Tests of a Damaged Ship in Head and Following Seas. In Proceedings of the 12th International Conference on the Stability of Ship and Ocean Vehicles, Glasgow, UK, 14–19 June 2015.

7. Siddiqui, M.A.; Greco, M.; Lugni, C.; Faltinsen, O.M. Experimental studies of a damaged ship section in forced heave motion. *Appl. Ocean. Res.* **2019**, *88*, 254–274. [CrossRef]

8. Rodrigues, J.M.; Lavrov, A.; Hinostroza, M.A.; Soares, C.G. Experimental and numerical investigation of the partial flooding of a barge model. *Ocean. Eng.* **2018**, *16*, 586–603. [CrossRef]

9. Korkut, E.; Altar, M.; Incecik, A. An experimental study of motion behavior with an intact and damaged Ro-Ro ship model. *Ocean. Eng.* **2004**, *31*, 483–512. [CrossRef]

10. Acanfora, M.; de Luca, F. An experimental investigation on the dynamic response of a damaged ship with a realistic arrangement of the flooded compartment. *Appl. Ocean. Res.* **2017**, *69*, 191–204. [CrossRef]

11. Domeh, V.D.K.; Sobey, A.J.; Hudson, D.A. A preliminary experimental investigation into the influence of compartment permeability on damaged ship response in waves. *Appl. Ocean. Res.* **2015**, *52*, 27–36. [CrossRef]

12. Risto, J.; Pekka, R.; Mateusz, W.; Hendrik, N.; Sander, V. A Study on Leakage and Collapse of Non-Watertight Ship Doors under Floodwater Pressure. *Mar. Struct.* **2017**, *51*, 188–201.

13. Gao, Z.; Vassalos, D.; Gao, Q. Numerical simulation of water flooding into a damaged vessel's compartment by the volume of fluid method. *Ocean. Eng.* **2010**, *37*, 1428–1442. [CrossRef]

14. Sadat-Hosseini, H.; Kim, D.H.; Carrica, P.M.; Rhee, S.H. URANS simulations for a flooded ship in calm and regular beam waves. *Ocean. Eng.* **2016**, *120*, 318–330. [CrossRef]

15. Santos Tiago, A.; Guedes Soares, C. Numerical assessment of factors affecting the survivability of damaged ro-ro ships in waves. *Ocean. Eng.* **2009**, *36*, 797–809. [CrossRef]

16. Ming, F.R.; Zhang, A.M.; Cheng, H.; Sun, P.N. Numerical simulation of a damaged ship cabin flooding in transversal waves with Smoothed Particle Hydrodynamics method. *Ocean. Eng.* **2018**, *165*, 336–352. [CrossRef]

17. Manderbacka, T.; Mikkola, T.; Ruponen, P.; Matusiak, J. Transient response of a ship to an abrupt flooding accounting for the momentum flux. *J. Fluids Struct.* **2015**, *57*, 108–126. [CrossRef]

18. Acanfora, M.; Begovic, E.; De Luca, F. A Fast Simulation Method for Damaged Ship Dynamics. *J. Mar. Sci. Eng.* **2019**, *7*, 111. [CrossRef]

19. Stern, F.; Yang, J.M.; Wang, Z.Y.; Sadat-Hosseini, H.; Mousaviraad, M.; Bhushan, S.; Xing, T. Computational ship hydrodynamics: Nowadays and way forward. *Int. Shipbuild. Prog.* **2013**, *60*, 3–105.

20. SIEMENS. STAR-CCM+ Overset Mesh. Available online: https://mdx2.plm.automation.siemens.com/sites/default/files/flier/pdf/Siemens-PLM-CD-adapco-STAR-CCM-overset-mesh-fs-59886-A2.pdf (accessed on 25 August 2019).

21. *STAR-CCM+ Users' Guide Version 12.02; CD-Adapco, Computational Dynamics-Analysis & Design*; Application Company Ltd.: Melville, NY, USA, 2012.

22. De Luca, F.; Mancini, S.; Miranda, S.; Pensa, C. An extended verification and validation study of CFD simulations for planing hulls. *J. Ship Res.* **2016**, *60*, 101–118. [CrossRef]

23. Lee, Y.; Chan, H.S.; Pu, Y.; Incecik, A.; Dow, R.S. Global Wave Loads on a Damaged Ship. *Ships Offshore Struct.* **2012**, *7*, 237–268. [CrossRef]

24. Cao, X.Y.; Ming, F.R.; Zhang, A.M. Multi-phase SPH modelling of air effect on the dynamic flooding of a damaged cabin. *Comput. Fluids* **2018**, *163*, 7–19. [CrossRef]

25. Ruponen, P.; Kurvinen, P.; Saisto, I.; Harras, J. Air compression in a flooded tank of a damaged ship. *Ocean. Eng.* **2013**, *57*, 64–71. [CrossRef]

26. Zhang, X.L.; Lin, Z.; Li, P.; Dong, Y.; Liu, F. Time domain simulation of damage flooding considering air compression characteristic. *Water* **2019**, *11*, 796. [CrossRef]

27. Gao, Z.; Wang, Y.; Su, Y.; Chen, L. Numerical study of damaged ship's compartment sinking with air compression effect. *Ocean. Eng.* **2018**, *147*, 68–76. [CrossRef]

28. IMO MSC. 362 (92). Revised Recommendations on a Standard Method for Evaluating Cross-Flooding Arrangements. Adopted on 14 June 2013. Available online: http://www.imo.org/en/KnowledgeCentre/IndexofIMOResolutions/Documents/MSC-MaritimeSafety/362(92).pdf (accessed on 23 December 2019).

29. Mancini, S.; Begovic, E.; Day, A.H.; Incecik, A. Verification and validation of numerical modelling of DTMB 5415 roll decay. *Ocean. Eng.* **2018**, *162*, 209–223. [CrossRef]

30. Handschel, S.; Köllisch, N.; Soproni, J.P.; Abdel-Maksoud, M. A numerical method for estimation of ship roll damping for large amplitudes. In Proceedings of the 29th Symposium on Naval Hydrodynamics, Gothenburg, Sweden, 26–31 August 2012.
31. ITTC. Practical Guidelines for Ship CFD Applications. In *ITTC Procedures and Guidelines*; 7.5-03-02-03; ITTC: Rio de Janerio, Brazil, 2011.
32. Tezdogan, T.; Demirel, Y.K.; Kellet, P.; Khorasanchi, M.; Incecik, A. Full-scale unsteady RANS CFD simulations of ship behavior and performance in head seas due to slow steaming. *Ocean. Eng.* **2015**, *97*, 186–206. [CrossRef]
33. Zhang, X.L.; Lin, Z.; Mancini, S.; Li, P.; Li, Z.; Liu, F. A Numerical Investigation on the Flooding Process of Multiple Compartments Based on the Volume of Fluid Method. *J. Mar. Sci. Eng.* **2019**, *7*, 211. [CrossRef]
34. Menter, F.R. Eddy Viscosity Transport Equations and Their Relation to the k-ε Model. *J. Fluids Eng.* **1997**, *119*, 876–884. [CrossRef]
35. Begovic, E.; Day, A.H.; Incecik, A.; Mancini, S. Roll damping assessment of intact and damaged ship by CFD and EFD methods. In Proceeding of the 12th International Conference on the Stability of Ships and Ocean Vehicles, Glasgow, UK, 14–19 June 2015.
36. Patankar, S.V.; Spalding, D.B. A calculation procedure for heat, mass and momentum transfer in three-dimensional parabolic flows. *Int. J. Heat Mass Transf.* **1972**, *15*, 1787–1806. [CrossRef]
37. Weiss, J.M.; Maruszewski, J.P.; Smith, W.A. Implicit solution of preconditioned Navier–Stokes equations using algebraic multigrid. *AIAA J.* **1999**, *37*, 29–36. [CrossRef]
38. Salim, S.M.; Cheah, S.C. Wall y+ Strategy for Dealing with Wall-bounded Turbulent Flows. In Proceedings of the International MultiConference of Engineers and Computer Scientists, Hong Kong, China, 18–20 March 2009.
39. Rodi, W. Experience with Two-Layer Models Combining the k-ε Model with a One-equation Model near the Wall. *AIAA Pap.* **1991**, 91-0216. [CrossRef]
40. Ruponen, R.; Queutey, P.; Kraskowski, M.; Jalonen, R.; Guilmineau, E. On the calculation of cross-flooding time. *Ocean. Eng.* **2012**, *40*, 27–39. [CrossRef]

Journal of
Marine Science and Engineering

Article

Numerical Study on Hydrodynamics of Ships with Forward Speed Based on Nonlinear Steady Wave

Tianlong Mei [1,2], Maxim Candries [2], Evert Lataire [2] and Zaojian Zou [1,3,*]

[1]   School of Naval Architecture, Ocean and Civil Engineering, Shanghai Jiao Tong University, Shanghai 200240, China; tackimay@sjtu.edu.cn

[2]   Maritime Technology Division, Ghent University, Technologiepark 60, 9052 Zwijnaarde, Belgium; maxim.candries@ugent.be (M.C.); evert.lataire@ugent.be (E.L.)

[3]   State Key Laboratory of Ocean Engineering, Shanghai Jiao Tong University, Shanghai 200240, China

*   Correspondence: zjzou@sjtu.edu.cn

Received: 31 December 2019; Accepted: 7 February 2020; Published: 10 February 2020

Abstract: In this paper, an improved potential flow model is proposed for the hydrodynamic analysis of ships advancing in waves. A desingularized Rankine panel method, which has been improved with the added effect of nonlinear steady wave-making (NSWM) flow in frequency domain, is employed for 3D diffraction and radiation problems. Non-uniform rational B-splines (NURBS) are used to describe the body and free surfaces. The NSWM potential is computed by linear superposition of the first-order and second-order steady wave-making potentials which are determined by solving the corresponding boundary value problems (BVPs). The so-called m_j terms in the body boundary condition of the radiation problem are evaluated with nonlinear steady flow. The free surface boundary conditions in the diffraction and radiation problems are also derived by considering nonlinear steady flow. To verify the improved model and the numerical method adopted in the present study, the nonlinear wave-making problem of a submerged moving sphere is first studied, and the computed results are compared with the analytical results of linear steady flow. Subsequently, the diffraction and radiation problems of a submerged moving sphere and a modified Wigley hull are solved. The numerical results of the wave exciting forces, added masses, and damping coefficients are compared with those obtained by using Neumann–Kelvin (NK) flow and double-body (DB) flow. A comparison of the results indicates that the improved model using the NSWM flow can generally give results in better agreement with the test data and other published results than those by using NK and DB flows, especially for the hydrodynamic coefficients in relatively low frequency ranges.

Keywords: nonlinear steady flow; desingularized Rankine panel method; forward speed; radiation and diffraction

---

1. Introduction

Over the last decades, the rapid development of computing power and the emergence of more sophisticated numerical methods have promoted the applications of numerical methods in ship hydrodynamics problems. Nevertheless, these problems still need to be simplified due to the complexity behind the physical models. It becomes even more complicated when different models need to be coupled, which is for example the case when ship maneuvering in waves is considered.

In the early stage, two-dimensional strip theory was developed as a practical way to evaluate ship hydrodynamic performances [1,2]. However, relying on the assumption that the ship is a slender body, strip theory is only suitable for low speed and high encounter wave frequency cases. In order to consider more realistic three-dimensional (3D) effects, it is not appropriate to assume that the ship is slender.

In the study of ship hydrodynamics problems, 3D potential flow theory has focused mainly on linear analysis [3]. The theory assumes that the disturbance due to the presence of a ship in waves is relatively small. When using Rankine panel methods, two linearization methods can be distinguished: the Neumann–Kelvin (NK) linearization and the double-body (DB) linearization. The former considers uniform flow as basic flow to linearize the free surface boundary conditions. The latter is essentially based on a slow-ship assumption which obtains the double-body velocity potential by treating the free surface as a rigid horizontal plane and takes the DB flow as basic flow to linearize the free surface boundary conditions. Numerous studies have been published using these two methods. For instance, Kim and Kim [3] presented a study on ship hydrodynamics comparing the NK and DB linearization methods. Similar researches discussing the advantages and disadvantages of these two methods can be found in Zhang et al. [4], Zhang and Beck [5], Zhang and Zou [6]. Attempts have also been made to include ship maneuvering in waves in the analysis, e.g., Seo and Kim [7], Zhang et al. [8]. In their studies, the mean second-order wave force was evaluated by Rankine panel method using NK or DB linearization, which was then treated as the input force in the equations for predicting maneuvering behavior. However, these two linearization methods, as described in [4], can be justified in the case of a slender ship, but they are not suitable for blunt bodies or ships moving at high speeds [9]. In light of the limitations of the NK and DB linearizations, works addressing ship hydrodynamics by using steady wave-making flow as basic flow for linearization were carried out; e.g., Gao and Zou [10] computed the linear steady wave-making flow beforehand and then applied the results to solve the diffraction and radiation problems. Recently, researchers have considered nonlinear steady flow to study the interactions between the linear periodic wave-induced flow and the nonlinear steady flow caused by the ship's forward speed in calm water, such as the studies by Bunnik [11], Söding et al. [12] and Chillcce and el Moctar [13] in frequency domain. As for the time domain method, studies can be found in Riesner et al. [14], Riesner and el Moctar [15] and Chen et al. [9]. Though the transient effect of flow can be investigated in time domain, the boundary integral equation should be solved at each time step, which is more computationally expensive than that with frequency domain method.

In this study, a new model is proposed to compute the ship hydrodynamic forces in the frequency domain. In contrast to other methods, nonlinear steady flow is considered and the interaction between nonlinear steady flow and unsteady flow is considered not only in the body boundary condition, but also in the free surface boundary conditions in the corresponding diffraction and radiation problems. The main objective of this method is to capture the coupling factors as accurately as possible. The boundary value problems (BVPs) for the first-order and second-order steady wave-making potentials are first derived and solved, and the nonlinear steady wave-making (NSWM) potential is then approximated by linear superposition of the first-order and second-order steady wave-making potentials. Subsequently, the wave exciting forces and the radiation forces are evaluated based on the obtained NSWM flow.

A desingularized Rankine panel method with distributed sources at a small distance inside the body and above the free surface [16,17] is applied to numerically solve the problems. This method has the advantage over conventional boundary integral methods in that it separates the integration surface and the collocation surface, which in turn results in a boundary integral equation with non-singular kernels. In addition, the second-order or even higher-order derivatives of the velocity potential can be directly evaluated without complicated numerical treatments to eliminate the singularities in the integral equation; thus, the method is faster and easier to implement. In recent years, this method has been extended and applied in the analysis of 2D wave-body interaction problems, such as Feng et al. [18–20].

To verify the proposed model, non-uniform rational B-splines (NURBS) are used to generate the mesh both on the body surface and the free surface. The desingularized Rankine panel method is then employed to discretize and solve the boundary integral equation. The m_j terms in the body boundary condition are evaluated with nonlinear steady flow, and the free surface boundary conditions in the diffraction and radiation problems are also derived by taking the nonlinear steady flow into account.

In order to identify the effects of steady flow on the unsteady flow, the present method is compared with those based on NK and DB linearizations. The computations are carried out for a submerged moving sphere and a modified Wigley hull advancing in head waves. The numerical results including the wave exciting forces, added masses and damping coefficients with the effects of different steady flows are presented.

2. Mathematical Formulations

Figure 1 shows the two coordinate systems that are used: an earth-fixed coordinate system $o_0 - x_0 y_0 z_0$ and a coordinate system $o - xyz$ moving along with the ship at a constant speed U with ox positive to the bow, oy positive to the port side and oz directing upwards.

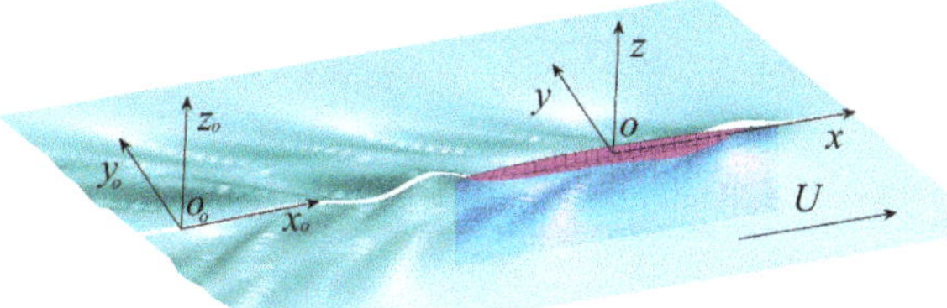

Figure 1. Coordinate systems.

In $o_0 - x_0 y_0 z_0$, based on the assumptions of ideal fluid and irrotational flow, the total velocity potential $\Psi(\vec{x}_0, t)$ should satisfy the following equations:

Laplace's equation in fluid domain:

$$\nabla^2 \Psi = 0 \tag{1}$$

The kinematic and dynamic boundary conditions on the free surface S_F:

$$\left(\frac{\partial}{\partial t} + \nabla \Psi \cdot \nabla\right)[z_0 - \eta(x_0, y_0, t)] = 0 \tag{2}$$

$$g\eta + \Psi_t + \frac{1}{2}\nabla \Psi \cdot \nabla \Psi = 0 \tag{3}$$

where η is the free surface elevation, g is the gravitational acceleration. The subscripts (i.e., t, x_0, y_0) denote the derivatives with respect to the corresponding variables.

By combining Equation (2) and Equation (3), the following boundary condition on S_F is derived:

$$\Psi_{tt} + \nabla \Psi \cdot \nabla \Psi_t + \Psi_{x_0} \Psi_{x_0 t} + \Psi_{y_0} \Psi_{y_0 t} + \Psi_{x_0} \nabla \Psi \cdot \nabla \Psi_{x_0} + \Psi_{y_0} \nabla \Psi \cdot \nabla \Psi_{y_0} + g\Psi_{z_0} = 0 \tag{4}$$

The boundary condition on the body surface S_B:

$$\nabla \Psi \cdot \vec{n} = \vec{V}_B \cdot \vec{n} \tag{5}$$

where $\vec{n}$ is the unit normal vector directed inward of the body surface with $(n_1, n_2, n_3) = \vec{n}, (n_4, n_5, n_6) = \vec{x} \times \vec{n}$; $\vec{V}_B$ is instantaneous velocity of the body surface S_B

Moreover, a radiation condition should be satisfied. The details for implementing the numerical radiation condition will be introduced in Section 3.

By using the Galilean transformation, the relation from $o_0 - x_0 y_0 z_0$ to $o - xyz$ can be transformed as:

$$\frac{d}{dt} = \frac{\partial}{\partial t} - U\frac{\partial}{\partial x} \tag{6}$$

where $\frac{d}{dt}$ is the time derivative in coordinate system $o_0 - x_0 y_0 z_0$ and $\frac{\partial}{\partial t}$ is the time derivative in the moving coordinate system $o - xyz$.

In $o - xyz$, assuming that the velocity potential $\Psi(\vec{x}, t)$ can be written as:

$$\Psi(\vec{x}, t) = -Ux + \Phi^S(\vec{x}) + \mathrm{Re}\left[A\varphi^I(\vec{x})e^{i\omega t} + A\varphi^D(\vec{x})e^{i\omega t}\right] + \mathrm{Re}\left\{\sum_{j=1}^{6}\left[\bar{\xi}_j\varphi_j^R(\vec{x})e^{i\omega t}\right]\right\} \tag{7}$$

where $\left[-Ux + \Phi^S(\vec{x})\right]$ is the steady velocity potential; $\varphi^I(\vec{x})$, $\varphi^D(\vec{x})$ and $\varphi_j^R(\vec{x})$ $(j = 1, 2, \cdots, 6)$ are the spatial parts of the incident, diffraction and radiation velocity potentials, respectively; A is the incoming wave amplitude, $\bar{\xi}_j$ $(j = 1, 2, \cdots, 6)$ is the amplitude of j-th mode of oscillating motion, and ω is the encounter frequency.

2.1. Nonlinear Steady Wave-Making (NSWM) Problem

Substituting Equation (7) into Equations (1)–(5), using $\Psi(x_0, y_0, z_0, t) = \Psi(x + Ut, y, z, t)$ and Equation (6), and extracting the terms unrelated to time t, the BVP of the steady wave-making velocity potential $\Phi^S(\vec{x})$ can be expressed in the moving coordinate system $o - xyz$ as:

Laplace's equation in fluid domain:

$$\nabla^2 \Phi^S = 0 \tag{8}$$

The boundary condition on the free surface S_F:

$$U^2\Phi_{xx}^S - UV\Phi^S \cdot \nabla\Phi_x^S - U\Phi_x^S \cdot \Phi_{xx}^S - U\Phi_y^S\Phi_{xy}^S + \Phi_x^S\nabla\Phi^S \cdot \nabla\Phi_x^S + \Phi_y^S\nabla\Phi^S \cdot \nabla\Phi_y^S + g\Phi_z^S = 0 \tag{9}$$

The boundary condition on the body surface S_B:

$$-Un_1 + \vec{n} \cdot \nabla\Phi^S = 0 \tag{10}$$

By using Equation (6), the steady hydrodynamic pressure can be obtained from Bernoulli's equation:

$$p^S = -\rho\left(\frac{1}{2}\nabla\Phi^S \cdot \nabla\Phi^S - U\Phi_x^S\right) \tag{11}$$

The steady force F_i^S $(i = 1, 2, \cdots, 6)$ can then be calculated by integrating the pressure over the wetted body surface:

$$F_i^S = \iint_{S_B} p^S n_i ds, \quad i = 1, 2, \cdots, 6 \tag{12}$$

By using Equation (6), the steady free surface elevation can be obtained from Equation (3):

$$\eta^S = \frac{U}{g}\Phi_x^S - \frac{1}{2g}\nabla\Phi^S \cdot \nabla\Phi^S \tag{13}$$

The boundary condition Equation (9) is nonlinear. To solve the resulting nonlinear BVP, the velocity potential Φ^S and the free surface elevation η^S are expressed by perturbation expansion until second order as:

$$\Phi^S \approx \Phi^{S(1)} + \Phi^{S(2)}$$
$$\eta^S \approx \eta^{S(1)} + \eta^{S(2)} \tag{14}$$

Substituting Equation (14) into Equations (8)–(10), the BVPs for the first- and second-order steady velocity potentials can be obtained by Taylor expansion on $z = 0$ and about the mean wetted body surface $\bar{S}_b$. The BVP for the first-order steady velocity potential is given as:

$$\nabla^2 \Phi^{S(1)} = 0, \text{ in fluid domain} \tag{15}$$

$$\frac{U^2}{g}\Phi_{xx}^{S(1)} + \Phi_z^{S(1)} = 0, \text{ on } z = 0 \tag{16}$$

$$\vec{n}\cdot\nabla\Phi^{S(1)} = Un_1, \text{ on the mean wetted body surface } \overline{S}_b \tag{17}$$

The BVP for the second-order steady velocity potential is given as:

$$\nabla^2\Phi^{S(2)} = 0, \text{ in fluid domain} \tag{18}$$

$$\begin{aligned}\frac{U^2}{g}\Phi_{xx}^{S(2)} + \Phi_z^{S(2)} = &-\frac{\partial}{\partial z}\left(\frac{U^2}{g}\Phi_{xx}^{S(1)} + \Phi_z^{S(1)}\right)\eta^{S(1)} + \frac{U}{g}\nabla\Phi_x^{S(1)}\cdot\nabla\Phi^{S(1)}\\ &+ \frac{U}{g}\left(\Phi_x^{S(1)}\Phi_{xx}^{S(1)} + \Phi_y^{S(1)}\Phi_{xy}^{S(1)}\right),\end{aligned} \qquad \text{on } z = 0 \tag{19}$$

$$\vec{n}\cdot\nabla\Phi^{S(2)} = 0, \text{ on the mean wetted body surface } \overline{S}_b \tag{20}$$

2.2. Diffraction Problem

For the diffraction problem, the ship moves with a constant speed U in waves without oscillations. In deep water, the incident wave velocity potential is given as:

$$\varphi^I(x,y,z) = \frac{ig}{\omega_0}e^{kz}\cdot e^{-ik(x\cos\beta + y\sin\beta)} \tag{21}$$

where ω_0 is the incident wave frequency, $k = \omega_0^2/g$ is the wave number, β is the wave angle and $\beta = \pi$ represents head sea condition. The encounter frequency ω is defined as:

$$\omega = \omega_0 - kU\cos\beta \tag{22}$$

Substituting Equation (7) into Equations (1)–(5), using $\Psi(x_0, y_0, z_0, t) = \Psi(x + Ut, y, z, t)$ and Equation (6), and extracting the terms related to time t and $\varphi^D(x,y,z)$, the BVP of the diffraction potential $\varphi^D(x,y,z)$ can be derived as:

Laplace's equation in fluid domain:

$$\nabla^2\varphi^D = 0 \tag{23}$$

The free surface boundary condition on $z = 0$:

$$\begin{aligned}&-\omega^2\varphi^D - 2i\omega U\varphi_x^D + U^2\varphi_{xx}^D + g\varphi_z^D + i\omega\nabla\Phi^S\cdot\nabla\varphi^D + i\omega\Phi_x^S\varphi_x^D + i\omega\Phi_y^S\varphi_y^D - UV\Phi^S\cdot\nabla\varphi_x^D\\ &-UV\varphi^D\cdot\nabla\Phi_x^S - U\Phi_x^S\varphi_{xx}^D + \Phi_x^S\nabla\Phi^S\cdot\nabla\varphi_x^D + \Phi_x^S\nabla\varphi^D\cdot\nabla\Phi_x^S - U\varphi_x^D\Phi_{xx}^S + \varphi_x^D\nabla\Phi^S\cdot\nabla\Phi_x^S\\ &-U\Phi_y^S\varphi_{xy}^D - U\varphi_y^D\Phi_{xy}^S + \Phi_y^S\nabla\Phi^S\cdot\nabla\varphi_y^D + \Phi_y^S\nabla\varphi^D\cdot\nabla\Phi_y^S + \varphi_y^D\nabla\Phi^S\cdot\nabla\Phi_y^S = RHS\end{aligned} \tag{24}$$

where $RHS = \omega^2\varphi^I + 2i\omega U\varphi_x^I - U^2\varphi_{xx}^I - g\varphi_z^I - i\omega\nabla\Phi^S\cdot\nabla\varphi^I - i\omega\Phi_x^S\varphi_x^I - i\omega\Phi_y^S\varphi_y^I + UV\Phi^S\cdot\nabla\varphi_x^I + UV\varphi^I\cdot\nabla\Phi_x^S + U\Phi_x^S\varphi_{xx}^I - \Phi_x^S\nabla\Phi^S\cdot\nabla\varphi_x^I - \Phi_x^S\nabla\varphi^I\cdot\nabla\Phi_x^S + U\varphi_x^I\Phi_{xx}^S - \varphi_x^I\nabla\Phi^S\cdot\nabla\Phi_x^S + U\Phi_y^S\varphi_{xy}^I + U\varphi_y^I\Phi_{xy}^S - \Phi_y^S\nabla\Phi^S\cdot\nabla\varphi_y^I - \Phi_y^S\nabla\varphi^I\cdot\nabla\Phi_y^S - \varphi_y^I\nabla\Phi^S\cdot\nabla\Phi_y^S$

The boundary condition on the mean wetted body surface $\overline{S}_b$:

$$\vec{n}\cdot\nabla\varphi^D = -\vec{n}\cdot\nabla\varphi^I \tag{25}$$

It is worth noting that in Equation (24), the nonlinear steady potential Φ^S is also considered in the free surface boundary condition.

Once the diffraction potential φ^D is obtained, the wave exciting forces on the hull can be computed as:

$$F_j = \text{Re}\left(Af_je^{i\omega t}\right), \ j = 1, 2, \cdots, 6 \tag{26}$$

$$f_j = -\rho\iint\limits_{\overline{S}_b}\left[i\omega(\varphi^I + \varphi^D) - U(\varphi_x^I + \varphi_x^D) + \nabla\Phi^S\cdot\nabla(\varphi^I + \varphi^D)\right]n_jds \tag{27}$$

2.3. Radiation Problem

For the radiation problem, it is assumed that the ship undergoes a harmonic oscillation. Similar to the BVP of diffraction potential, the radiation potential φ_j^R should satisfy the control equation and boundary conditions below:

Laplace's equation in fluid domain:

$$\nabla^2 \varphi_j^R = 0, j = 1, 2, \cdots, 6 \tag{28}$$

The free surface boundary condition on $z = 0$:

$$\begin{aligned}
&-\omega^2 \varphi_j^R - 2i\omega U \varphi_{jx}^R + U^2 \varphi_{jxx}^R + g\varphi_{jz}^R + i\omega \nabla \Phi^S \cdot \nabla \varphi_j^R + i\omega \Phi_x^S \varphi_{jx}^R + i\omega \Phi_y^S \varphi_{jy}^R \\
&-UV\Phi^S \cdot \nabla \varphi_{jx}^R - UV\varphi_j^R \cdot \nabla \Phi_x^S - U\Phi_x^S \varphi_{jxx}^R + \Phi_x^S \nabla \Phi^S \cdot \nabla \varphi_{jx}^R + \Phi_x^S \nabla \varphi_j^R \cdot \nabla \Phi_x^S \\
&-U\varphi_{jx}^R \Phi_{xx}^S + \varphi_{jx}^R \nabla \Phi^S \cdot \nabla \Phi_x^S - U\Phi_y^S \varphi_{jxy}^R - U\varphi_{jy}^R \Phi_{xy}^S + \Phi_y^S \nabla \Phi^S \cdot \nabla \varphi_{jy}^R \\
&+\Phi_y^S \nabla \varphi_j^R \cdot \nabla \Phi_y^S + \varphi_{jy}^R \nabla \Phi^S \cdot \nabla \Phi_y^S = 0
\end{aligned} \tag{29}$$

The boundary condition on the mean wetted body surface $\overline{S}_b$:

$$\vec{n} \cdot \nabla \varphi_j^R = -i\omega n_j + m_j \tag{30}$$

where the m_j terms representing the coupling effect between the steady and unsteady flows are given as:

$$\begin{aligned}
(m_1, m_2, m_3) &= (\vec{n} \cdot \nabla)(\vec{U} - \nabla \Phi^S) \\
(m_4, m_5, m_6) &= (\vec{n} \cdot \nabla)\left[\vec{x} \times (\vec{U} - \nabla \Phi^S)\right]
\end{aligned} \tag{31}$$

where $\vec{U} = (U, 0, 0)$.

It is worth noting that the effect of the nonlinear steady potential Φ^S occurs not only in the so-called m_j terms in Equation (30), but also in the free surface boundary condition Equation (29).

Once the radiation potential φ_j^R is determined, the added mass a_{kj} and damping coefficient b_{kj} $(k, j = 1, 2, \cdots, 6)$ can be obtained as:

$$\begin{aligned}
a_{kj} &= \frac{-\rho}{\omega^2} \mathrm{Re} \iint_{\overline{S}_b} (i\omega \varphi_j^R - U\varphi_{jx}^R + \nabla \Phi^S \cdot \nabla \varphi_j^R) n_k ds \\
b_{kj} &= \frac{-\rho}{\omega} \mathrm{Im} \iint_{\overline{S}_b} (i\omega \varphi_j^R - U\varphi_{jx}^R + \nabla \Phi^S \cdot \nabla \varphi_j^R) n_k ds
\end{aligned} \tag{32}$$

3. Desingularized Rankine Panel Method

In this paper, a desingularized Rankine panel method is applied, where the Rankine sources are distributed inside the body and above the free surface at a distance L_d according to the formula $L_d = l_d(D_m)^v$ proposed by Cao et al. [17], l_d and v are equal to 1.0 and 0.5 respectively and D_m is the local mesh size (the square root of the local mesh area), as demonstrated in Figure 2.

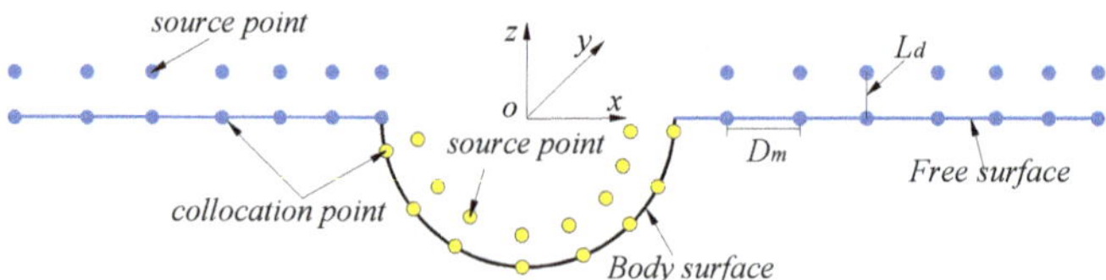

Figure 2. Desingularized Rankine panel model.

A suitable radiation condition should be implemented to ensure a unique solution for the specific BVP when using the Rankine source method. Typically, the numerical techniques can be classified as follows:

- The upstream radiation condition [21]: imposing a boundary condition by difference method at the upstream boundary of the truncated free surface to ensure that no scattered waves propagate ahead of the vessel.
- The staggered method [22]: shifting the source points above the free surface a certain distance downstream.
- The fluid domain decomposition method [23]: the flow field is divided into an inner domain and an outer domain by vertical control surfaces, where the Rankine source is adopted in the inner area, while the Kelvin source is adopted in the outer domain, and the solutions are matched on the control surfaces.
- The modified Sommerfeld radiation condition [24,25]: the modified Sommerfeld radiation condition is adopted by taking account the Doppler shift of the scattered waves at the control surface that truncates the infinite fluid domain.

In this study, the radiation condition is satisfied by using the staggered method for its simple implementation and good stability. The raised source points are moved a distance Δx toward downstream. The recommended parameter in this study is $\Delta x = \delta$, where δ denotes the average longitudinal value between two adjacent collocation points on the free surface. However, it should be noted that this numerical treatment is only valid for steady wave-making problem and the radiation problem with the Brard number $\tau = U\omega/g > 0.25$.

By using NURBS, the points (x, y, z) on the body and free surfaces can be described with parameter coordinate (u, v) as:

$$[x(u,v), y(u,v), z(u,v)] = \left[\sum_{i=0}^{m}\sum_{j=0}^{n} \omega_{ij}D_{ij}N_{i,k}(u)N_{j,l}(v)\right] \bigg/ \left[\sum_{i=0}^{m}\sum_{j=0}^{n} \omega_{ij}N_{i,k}(u)N_{j,l}(v)\right] \tag{33}$$

where D_{ij} are the control points on the body and free surfaces; ω_{ij} is the weight; $N_{i,k}(u)$ and $N_{j,l}(v)$ are the B-spline basis functions of $k(l)$-th order for a given knot sequence $u = (u_0, u_1, \cdots, u_{n+k+1})$, defined as:

$$\begin{cases} N_{i,0}(u) = \begin{cases} 1, & u_i \le u < u_{i+1} \\ 0, & \text{otherwise} \end{cases} \\ N_{i,k}(u) = \frac{u-u_i}{u_{i+k}-u_i}N_{i,k-1}(u) + \frac{u_{i+k+1}-u}{u_{i+k+1}-u_{i+1}}N_{i+1,k-1}(u) \end{cases} \tag{34}$$

According to Green's theorem, the velocity potential $\varphi(P)$ in the flow field can be expressed as:

$$\varphi(P) = \iint_S \sigma(Q)\frac{1}{r_{PQ}}dS = \iint_S \sigma(\vec{x}_s)\frac{1}{\left|\vec{x}_c - \vec{x}_s\right|}dS \tag{35}$$

where P is the field point, Q is the source point on the integration surface S, r_{PQ} represents the distance between the field point and the source point; $\sigma(Q)$ is the source strength distribution over the surface S. $\vec{x}_c$ and $\vec{x}_s$ represent the coordinates of collocation point and source point, respectively.

Applying the corresponding boundary conditions on the free surface Γ_f and body surface Γ_b, the integral equations for the unknown source strengths can be established and solved. The velocity potential on Γ_f and the normal derivative of the velocity potential on Γ_b are calculated by:

$$\iint_{S_f} \sigma(\vec{x}_s^f)\frac{1}{\left|\vec{x}_c^f - \vec{x}_s^f\right|}ds + \iint_{S_b} \sigma(\vec{x}_s^b)\frac{1}{\left|\vec{x}_c^f - \vec{x}_s^b\right|}ds = \varphi_0(\vec{x}_c^f), \ \vec{x}_c^f \in \Gamma_f \tag{36}$$

$$\iint_{S_f} \sigma(\vec{x}_s^f) \frac{\partial}{\partial n} \left(\frac{1}{\left| \vec{x}_c^b - \vec{x}_s^f \right|} \right) ds + \iint_{S_b} \sigma(\vec{x}_s^b) \frac{\partial}{\partial n} \left(\frac{1}{\left| \vec{x}_c^b - \vec{x}_s^b \right|} \right) dS_b = \frac{\partial}{\partial n} \varphi_0(\vec{x}_c^b), \ \vec{x}_c^b \in \Gamma_b \tag{37}$$

where $\vec{x}_c^f$ and $\vec{x}_c^b$ denote the collocation points on the free surface Γ_f and the body surface Γ_b; $\vec{x}_s^f$ and $\vec{x}_s^b$ denote the source points on the integration surface. S_f is the integration surface above the free surface Γ_f, and S_b is the integration surface inside the body surface Γ_b. $\varphi_0(\vec{x}_c^f)$ is the given velocity potential at $\vec{x}_c^f$ and $\frac{\partial}{\partial n} \varphi_0(\vec{x}_c^b)$ is the given normal derivative of the velocity potential at $\vec{x}_c^b$.

Discretizing the body surface and the free surface into N_b and N_f quadrilateral panels respectively, a set of discrete equations can be obtained from the integral equations. From Equations (36) and (37) it follows:

$$\sum_{j=1}^{N_f} \sigma_j^f(\vec{x}_s^f) \frac{1}{\left| \vec{x}_{ci}^f - \vec{x}_{sj}^f \right|} + \sum_{j=1}^{N_b} \sigma_j^b(\vec{x}_s^b) \frac{1}{\left| \vec{x}_{ci}^f - \vec{x}_{sj}^b \right|} = \varphi_0(\vec{x}_{ci}^f) \, , \ i = 1, 2, \cdots, N_f \tag{38}$$

$$\sum_{j=1}^{N_f} \sigma_j^f(\vec{x}_s^f) \frac{\partial}{\partial n_i} \left(\frac{1}{\left| \vec{x}_{ci}^b - \vec{x}_{sj}^f \right|} \right) + \sum_{j=1}^{N_b} \sigma_j^b(\vec{x}_s^b) \frac{\partial}{\partial n_i} \left(\frac{1}{\left| \vec{x}_{ci}^b - \vec{x}_{sj}^b \right|} \right) = \frac{\partial}{\partial n_i} \varphi_0(\vec{x}_{ci}^b) \, , \ i = 1, 2, \cdots, N_b \tag{39}$$

As can be seen from the discrete equations of Equations (38) and (39), the total number of equations is equal to the number of unknowns, i.e., $N = N_b + N_f$. Therefore, by satisfying the corresponding boundary conditions on the body surface and free surface at the collocation points, a set of linear equations for the unknown source strengths can be obtained. By solving these equations, the source strengths can be determined.

4. Numerical Results and Discussion

Two cases are studied: a sphere given by Equation (40), and a Wigley I ship [26] given by Equation (41):

$$x^2 + y^2 + (z - h)^2 = r^2 \tag{40}$$

$$y = \frac{B}{2} \left\{ \left[1 - \left(\frac{z}{T} \right)^2 \right] \left[1 - \left(\frac{2x}{L} \right)^2 \right] \left[1 + 0.2 \left(\frac{2x}{L} \right)^2 \right] + \left(\frac{z}{T} \right)^2 \left[1 - \left(\frac{z}{T} \right)^8 \right] \left[1 - \left(\frac{2x}{L} \right)^2 \right]^4 \right\} \tag{41}$$

where r is the radius of the sphere and h is the submerged depth; L, B and T are the length, the beam and the draft of the hull respectively. The Wigley I ship has the length to beam ratio $L/B = 10$ and the beam to draft ratio $B/T = 1.6$.

Figure 3 shows the typical panel arrangements of the submerged sphere and the Wigley I ship. In addition, the panel arrangements on the raised plane (cyan) above the free surface are shown. In Figure 3a, the free surface of the computational domain extends to $5.0r$ upstream, $5.0r$ sideways and $10.0r$ downstream. The discretized panels of the sphere and the half width free surface are 21×21 and 50×16, respectively. In Figure 3b, the free surface of the computational domain extends to $1.0L$ upstream, $0.75L$ sideways and $1.5L$ downstream. The discretized panels of the half Wigley I ship and half width free surface are 30×10 and 76×19, respectively.

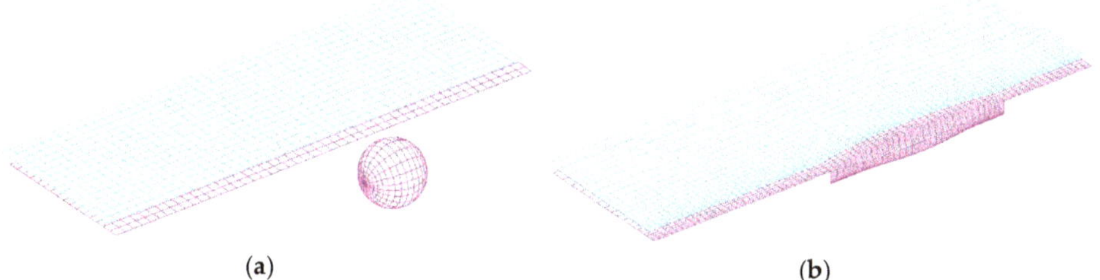

(a)   (b)

Figure 3. Typical panel arrangements of (**a**) submerged sphere and (**b**) Wigley I ship.

4.1. Results of the NSWM Problem

In order to compute the NSWM flow, a submerged moving sphere of $r = 1.0$ m at three different submerged depths ($h = 1.5r$, $2.0r$, $3.0r$) is chosen as the study case, the Froude number is defined as $Fr = U/\sqrt{gh}$, so the results at the same Froude number will correspond to different forward speeds when the submerged depth is varied. The numerical results are compared with the analytical results of Wu and Taylor [27]. Figure 4 shows the dimensionless nonlinear wave-making resistance coefficient C_w and lift force coefficient C_L of the sphere, where the "linear" results are obtained by solving the BVP of the first-order steady velocity potential, the "nonlinear" results are obtained from the BVPs of the superposition of the first-order and second-order steady velocity potentials; $C_w = -F_1^S/(\rho g \pi\, r^3)$ and $C_L = F_3^S/(\rho g \pi\, r^3)$.

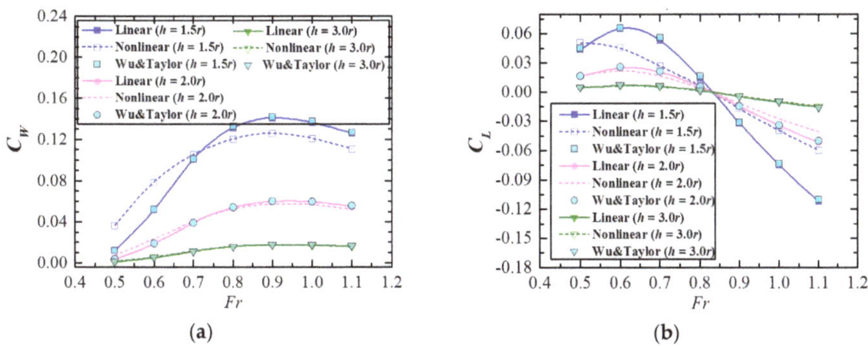

(a)   (b)

Figure 4. Coefficients of wave-making resistance (**a**) and lift force (**b**) at different *Fr*.

As can be seen from Figure 4, the present results are in good agreement with the analytical results in Wu and Taylor [27] for the linear solution. As can be seen in Figure 4a, the nonlinear results are larger than the linear results when the Froude number is less than a certain threshold value, whereas the situation reverses when it exceeds the threshold. However, a converse trend can be seen for the lift force in Figure 4b. A similar result can also be found in Kim [28]. This may be attributed to the "bow and stern wave-making effect", i.e., when a nonlinear free surface boundary condition is considered, the pressures at the bow and the stern will be different from those when a linear free surface boundary condition is considered. In addition, the differences between the linear and nonlinear results decrease when the submerged depth increases, which demonstrates that the effect of the nonlinear boundary condition on the free surface can be ignored when the submerged depth exceeds a certain value, which is also the case in reality.

4.2. m_j-Terms

The difficulty in solving a radiation problem lies in the accurate calculation of m_j terms, which contain the second-order derivatives of the steady velocity potential in the body boundary condition [29]. In order to calculate the m_j-terms to verify the calculation accuracy of the derivatives of the velocity potential on the body surface, the desingularized method is applied to a sphere ($r = 1.0$ m) moving at a speed $U = 1.0$ m/s in unbounded fluid.

The results of the first-order and second-order derivatives of the velocity potential are shown in Figure 5a–c, and the results of m_1, m_2 are shown in Figure 5d. It shows that the numerical results virtually coincide with the analytical solutions, which demonstrates that the present method is suitable for calculating the first- and second-order derivatives of the velocity potential on the body surface.

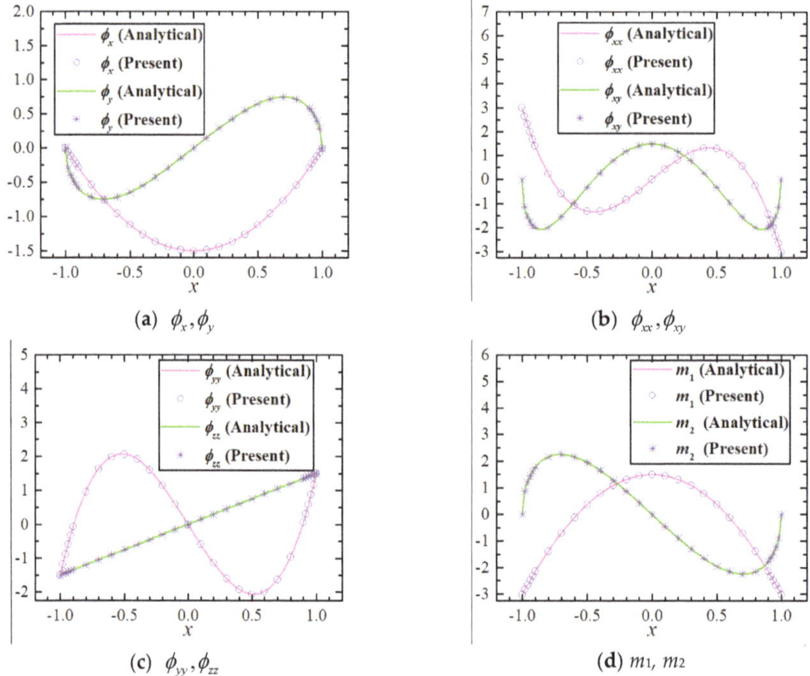

Figure 5. Derivatives of the velocity potential and m_j terms on the sphere surface along equator line.

4.3. Results of the Diffraction Problem

Tables 1 and 2 present the non-dimensional real and imaginary parts of the surge and heave wave exciting forces on the submerged sphere ($h = 2.0r$) moving at $Fr = U/\sqrt{gr} = 0.4$ in head waves as function of the non-dimensional wave number obtained by the present method in comparison with the analytical results in [27], where $k_e = \omega^2/g$. As can be seen in Tables 1 and 2, in general the present results based on the NSWM flow are in better agreement with those in [27] than the results based on the NK flow. Some deviations are observed at low frequencies, especially for the results based on the NK flow. There are two explanations for this larger deviation at low frequencies: firstly, NK flow cannot deal accurately with the relatively high forward speed because the wave disturbance induced by the forward speed of the body is neglected. Secondly, there exists a critical frequency $k_c r$ at the Brard number $\tau = U\omega/g = 0.25$, which is associated with the radiation condition [10]. Since the critical frequency is $k_c r = 0.2608$ for this case, poor accuracy is resulted when the frequency is near the critical frequency.

Table 1. Surge wave exciting forces on the sphere ($Fr = 0.4$, $h = 2.0r$, $kcr = 0.26808$).

kr	$f_1/(\rho g \pi A r^3 k_e)$ (Real Part)			$f_1/(\rho g \pi A r^3 k_e)$ (Imaginary Part)		
	Analytical Results [27]	Present (Nonlinear Steady Wave-Making, NSWM)	Present (Neumann–Kelvin, NK)	Analytical Results [27]	Present (NSWM)	Present (NK)
0.4	−0.0081	−0.00898	−0.00973	−0.5827	−0.58452	−0.61944
0.5	−0.0102	−0.01025	−0.01121	−0.4524	−0.45413	−0.47710
0.6	−0.0098	−0.00975	−0.00997	−0.3525	−0.35387	−0.35335
0.7	−0.0082	−0.00826	−0.00860	−0.2758	−0.27684	−0.27274
0.8	−0.0065	−0.00644	−0.00668	−0.2166	−0.21749	−0.21142
0.9	−0.0049	−0.00487	−0.00513	−0.1707	−0.17146	−0.16897
1.0	−0.0036	−0.00356	−0.00333	−0.135	−0.13561	−0.13653
1.2	−0.0018	−0.00182	−0.00199	−0.0851	−0.08553	−0.08533
1.4	−0.0009	−0.00089	−0.00100	−0.0541	−0.05436	−0.05635
1.6	−0.0004	−0.00042	−0.00035	−0.0346	−0.03476	−0.03859
1.8	−0.0002	−0.00020	−0.00020	−0.0222	−0.02233	−0.02638
2.0	−0.0001	−0.00012	−0.00012	−0.0143	−0.01443	−0.01533

Table 2. Heave wave exciting forces on the sphere ($Fr = 0.4$, $h = 2.0r$, $k_c r = 0.26808$).

kr	$f_3/(\rho g \pi A r^3 k_e)$ (Real Part)			$f_3/(\rho g \pi A r^3 k_e)$ (Imaginary Part)		
	Analytical Results [27]	Present (NSWM)	Present (NK)	Analytical Results [27]	Present (NSWM)	Present (NK)
0.4	−0.5691	−0.56878	−0.70940	0.0310	0.02993	0.03530
0.5	−0.4380	−0.43853	−0.41167	0.0248	0.02518	0.02611
0.6	−0.3398	−0.33971	−0.34704	0.0187	0.01872	0.01980
0.7	−0.2653	−0.26541	−0.27041	0.0136	0.01344	0.01479
0.8	−0.2082	−0.20856	−0.21633	0.0097	0.00977	0.01051
0.9	−0.1642	−0.16431	−0.17955	0.0069	0.00698	0.00727
1.0	−0.1299	−0.12995	−0.15140	0.0048	0.00483	0.00506
1.2	−0.0820	−0.08205	−0.09850	0.0023	0.00224	0.00239
1.4	−0.0522	−0.05224	−0.05490	0.0011	0.00107	0.00158
1.6	−0.0334	−0.03340	−0.03710	0.0005	0.00050	0.00052
1.8	−0.0214	−0.02146	−0.02938	0.0002	0.00021	0.00021
2.0	−0.0138	−0.01383	−0.01844	0.0001	0.00010	0.00017

Figure 6 shows the non−dimensional amplitudes and corresponding phase angles of heave and pitch wave exciting force/moment for the Wigley I ship advancing at $Fr = U/\sqrt{gL} = 0.4$ in head waves, where ∇ is the displacement volume. In order to investigate the influence of different steady flow models on wave exciting forces at various wave frequencies, the results based on the NK, DB and NSWM flows are compared in Figure 6. From Figure 6 one can observe that the present results obtained based on the three different steady flow models are in favourable agreement with the results obtained by Kara and Vassalos [30] using a 3D time domain method based on a transient free surface Green function, as well as with the experimental results by Journée [26]. In addition, one can also find that the results based on the NSWM flow and other two methods based on the NK and DB flows do not show evident differences, the reasons can be explained as follows: on one hand, though the effect of nonlinear steady flow is considered in the free surface boundary condition Equation (24), the interaction has no relation with the diffraction potential in the body surface boundary condition Equation (25); on the other hand, the small differences can be attributed to the predominant proportion of the Froude–Krylov force in the wave exciting force. Therefore, it can be concluded that the effect of the NSWM potential contributes unremarkably to the wave exciting force.

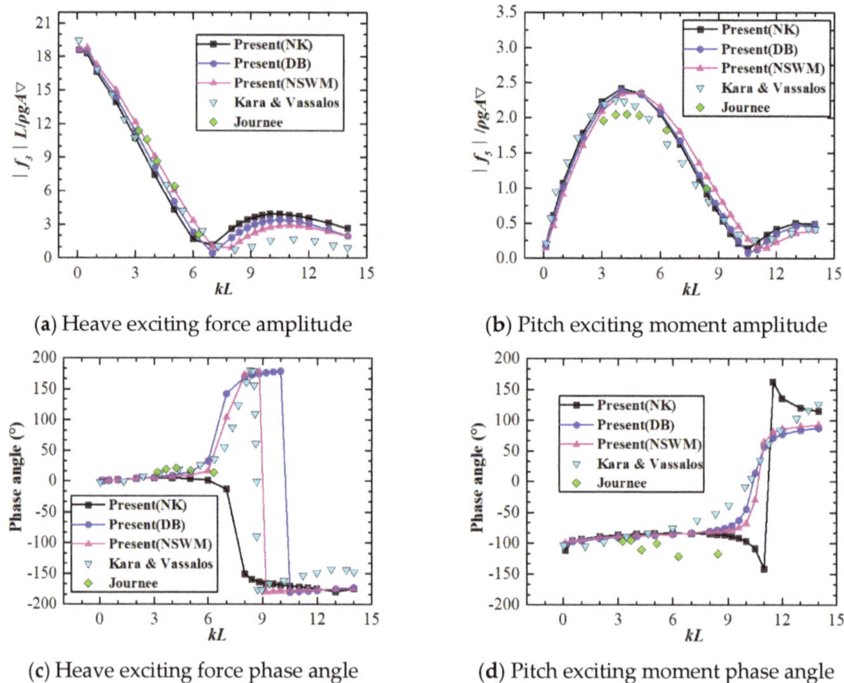

(a) Heave exciting force amplitude

(b) Pitch exciting moment amplitude

(c) Heave exciting force phase angle

(d) Pitch exciting moment phase angle

Figure 6. Amplitudes and phase angles of heave and pitch exciting force/moment on Wigley I ship (*Fr* = 0.4).

Figure 7 shows the real part of the diffraction wave contour for the submerged sphere (*Fr* = 0.4, *kr* = 0.5) and the Wigley I ship (*Fr* = 0.4, *kL* = 2π) based on the NSWM flow.

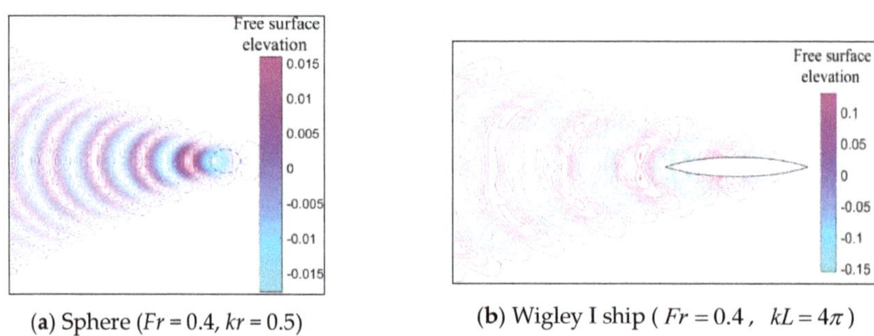

(a) Sphere (*Fr* = 0.4, *kr* = 0.5)

(b) Wigley I ship (*Fr* = 0.4 , *kL* = 4π)

Figure 7. Real part of diffraction wave contour of (**a**) submerged sphere and (**b**) Wigley I ship.

4.4. Results of the Radiation Problem

Tables 3–5 present the added masses and damping coefficients of the submerged sphere (*h* = 2.0*r*) moving at $Fr = U/\sqrt{gr} = 0.4$ in surge, sway and heave motions respectively, where the added masses and damping coefficients are non-dimensionalized as $A_{ij} = a_{ij}/(\pi\rho r^3)$, $B_{ij} = b_{ij}/(\pi\rho\omega r^3)$, $i,j = 1,2,3$. The $k_{ec}r$ corresponds to the critical frequency ω_c at the Brard number $\tau = 0.25$. As it can be seen from these tables, the numerical results based on the NSWM flow agree well with the analytical results in [27] and the numerical results in [10]. It should be noted that the linear steady wave-making potential Φ^S

was used to evaluate m_j terms in [10], without considering Φ^S in the free surface boundary condition. From these results, it can be concluded that the effects of the free surface nonlinearities are very weak due to the submerged depth.

Table 3. Added masses and damping coefficients of the sphere in surge motion ($Fr = 0.4$, $h = 2.0r$, $k_{ec}r = 0.3906$).

$k_e r$	A_{11}			B_{11}		
	Analytical Results [27]	Numerical Results [10]	Present (NSWM)	Analytical Results [27]	Numerical Results [10]	Present (NSWM)
0.6	1.2378	1.2358	1.25532	0.0362	0.0365	0.03447
0.7	1.1615	1.1614	1.16928	0.0247	0.0250	0.02586
0.8	1.1021	1.1021	1.10996	0.0195	0.0197	0.01972
0.9	1.0545	1.0544	1.06164	0.0169	0.0170	0.01724
1.0	1.0154	1.0153	1.02189	0.0154	0.0155	0.01572
1.5	0.8934	0.8933	0.89818	0.0113	0.0113	0.01144
2.0	0.8310	0.8310	0.83495	0.0079	0.0079	0.00794
2.5	0.7941	0.7940	0.79751	0.0052	0.0052	0.00519
3.0	0.7699	0.7699	0.77304	0.0033	0.0033	0.00327
3.5	0.7529	0.7529	0.75584	0.0021	0.0021	0.00207
4.0	0.7403	0.7403	0.74306	0.0013	0.0013	0.00131
4.5	0.7306	0.7305	0.73318	0.0008	0.0008	0.00081
5.0	0.7228	0.7228	0.72529	0.0005	0.0005	0.00050

Table 4. Added masses and damping coefficients of the sphere in sway motion ($Fr = 0.4$, $h = 2.0r$, $k_{ec}r = 0.3906$).

kr	A_{22}			B_{22}		
	Analytical Results [27]	Numerical Results [10]	Present (NSWM)	Analytical Results [27]	Numerical Results [10]	Present (NSWM)
0.6	1.1330	1.1357	1.13521	0.0771	0.0769	0.09103
0.7	1.0517	1.0528	1.05084	0.0650	0.0655	0.06482
0.8	0.9933	0.9946	0.99414	0.0544	0.0551	0.05481
0.9	0.9406	0.9506	0.95030	0.0454	0.0459	0.04605
1.0	0.9159	0.9167	0.91640	0.0380	0.0385	0.03851
1.5	0.8215	0.8219	0.82204	0.0164	0.0166	0.01656
2.0	0.7783	0.7787	0.77908	0.0076	0.0076	0.00765
2.5	0.7535	0.7538	0.75435	0.0037	0.0037	0.00368
3.0	0.7371	0.7374	0.73805	0.0019	0.0019	0.00184
3.5	0.7254	0.7256	0.72638	0.0010	0.0010	0.00095
4.0	0.7165	0.7168	0.71757	0.0005	0.0005	0.00051
4.5	0.7095	0.7098	0.71066	0.0003	0.0003	0.00027
5.0	0.7040	0.7042	0.70507	0.0002	0.0002	0.00015

Table 5. Added masses and damping coefficients of the sphere in heave motion ($Fr = 0.4$, $h = 2.0r$, $k_{ec}r = 0.3906$).

	A_{33}			B_{33}		
kr	Analytical Results [27]	Numerical Results [10]	Present (NSWM)	Analytical Results [27]	Numerical Results [10]	Present (NSWM)
0.6	1.0569	1.0526	1.05860	0.1090	0.1119	0.09671
0.7	0.9928	0.9925	0.98991	0.0862	0.0874	0.08827
0.8	0.9449	0.9447	0.94204	0.0709	0.0712	0.07146
0.9	0.9076	0.9076	0.90508	0.0597	0.0606	0.06026
1.0	0.8780	0.8775	0.87549	0.0511	0.0606	0.05162
1.5	0.7912	0.7911	0.78990	0.0264	0.0261	0.02659
2.0	0.7504	0.7503	0.74954	0.0145	0.0146	0.01473
2.5	0.7272	0.7271	0.72632	0.0083	0.0083	0.00828
3.0	0.7123	0.7123	0.71135	0.0048	0.0048	0.00478
3.5	0.7020	0.7021	0.70125	0.0028	0.0029	0.00276
4.0	0.6942	0.6942	0.69367	0.0017	0.0017	0.00169
4.5	0.6882	0.6882	0.68768	0.0010	0.0010	0.00100
5.0	0.6833	0.6833	0.68286	0.0006	0.0006	0.00064

Table 6 presents the coupling added masses and coefficients. It can be seen that the present results almost show the reverse relations, i.e., $(A_{13}, B_{13}) = (-A_{31}, -B_{31})$, which is consistent with the results in [27].

Table 6. Coupling added masses and damping coefficients of the sphere ($Fr = 0.4$, $h = 2.0r$, $k_{ec}r = 0.3906$).

	A_{13}		A_{31}		B_{13}		B_{31}	
kr	Analy. [27]	Present (NSWM)	Analy. [27]	Present (NSWM)	Analy. [27]	Present (NSWM)	Analy. [27]	Present (NSWM)
0.6	−0.0269	−0.02001	0.0269	0.01927	−0.0985	−0.09120	0.0985	0.08952
0.7	−0.0075	−0.00918	0.0075	0.00840	−0.0791	−0.07988	0.0791	0.07865
0.8	0.0030	0.00331	−0.0030	−0.00356	−0.0642	−0.06466	0.0642	0.06351
0.9	0.0089	0.00930	−0.0089	−0.00935	−0.0527	−0.05271	0.0527	0.05193
1.0	0.0122	0.01235	−0.0122	−0.01253	−0.0437	−0.04345	0.0437	0.04291
1.5	0.0135	0.01329	−0.0135	−0.01360	−0.0187	−0.01880	0.0187	0.01818
2.0	0.0095	0.00958	−0.0095	−0.00960	−0.0093	−0.00954	0.0093	0.00885
2.5	0.0061	0.00624	−0.0061	−0.00602	−0.0056	−0.00567	0.0056	0.00533
3.0	0.0038	0.00377	−0.0038	−0.00377	−0.0042	−0.00412	0.0042	0.00385
3.5	0.0023	0.00226	−0.0023	−0.00229	−0.0036	−0.00365	0.0036	0.00341
4.0	0.0014	0.00141	−0.0014	−0.00145	−0.0034	−0.00349	0.0034	0.00321
4.5	0.0009	0.00089	−0.0009	−0.00085	−0.0034	−0.00341	0.0034	0.00315
5.0	0.0006	0.00055	−0.0006	−0.00054	−0.0033	−0.00334	0.0033	0.00315

Figure 8 shows the heave and pitch hydrodynamic coefficients of the Wigley I ship at $Fr = U/\sqrt{gL} = 0.4$, where the coupling hydrodynamic coefficients are non-dimensionalized as $A_{35(53)} = a_{35(53)}/(\rho \nabla L)$, $B_{35(53)} = b_{35(53)}/(\rho \nabla L \sqrt{g/L})$. As it can be seen in Figure 8, good agreement is achieved among the present numerical results and the results in [30] using NK flow and a transient free surface Green function method, and the experimental results by Journée [26]. The results based on the NSWM flow show in general better agreement with the experimental results than those obtained using DB and NK flows, especially in the low frequency ranges. However, a remarkable deviation can be observed for the heave damping coefficient in Figure 8b. This is because the uniform flow is taken as the basic flow in the method using NK flow, correspondingly the second-order derivatives of the steady potential Φ^S are neglected in the m_j terms. As a result, this treatment cannot accurately reflect the interaction between the steady flow and the unsteady flow in the body surface boundary condition. On the other hand, as explained in [9], the larger contribution of the steady velocity potential in both the body surface and free surface boundary conditions at low frequencies also leads to relatively larger deviations, but these effects are not fully considered in the method using DB flow. Therefore, the coupling effects between the nonlinear steady flow and the unsteady flow, which are reflected in both the free surface boundary condition and the m_j terms in the body surface boundary condition, are quite important for the prediction of hydrodynamic coefficients, especially at low frequencies.

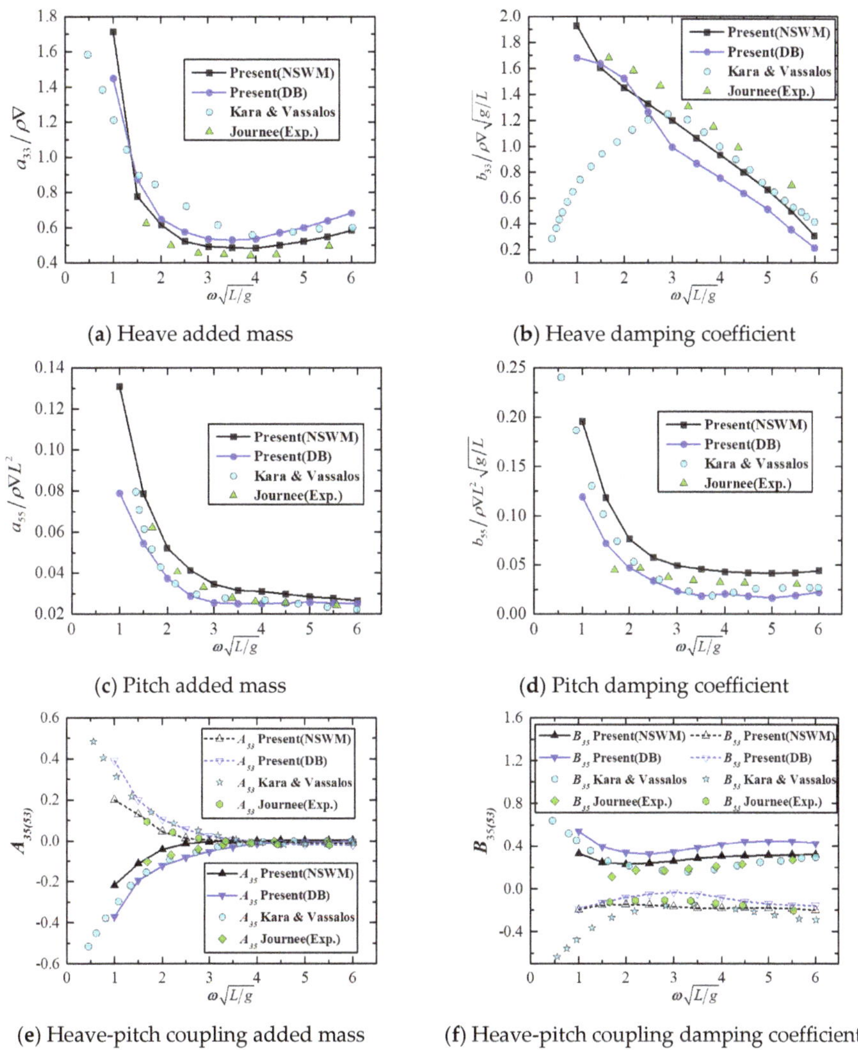

(**a**) Heave added mass

(**b**) Heave damping coefficient

(**c**) Pitch added mass

(**d**) Pitch damping coefficient

(**e**) Heave-pitch coupling added mass

(**f**) Heave-pitch coupling damping coefficient

Figure 8. Added masses and damping coefficients of Wigley I ship ($Fr = 0.4$).

Figure 9 shows the real part of the heave radiation wave contour of the sphere ($Fr = U/\sqrt{gr} = 0.4$, $k_e r = 2.0$) and the Wigley I ship ($Fr = 0.4$, $\omega/\sqrt{g/L} = 3.0$) based on the NSWM flow.

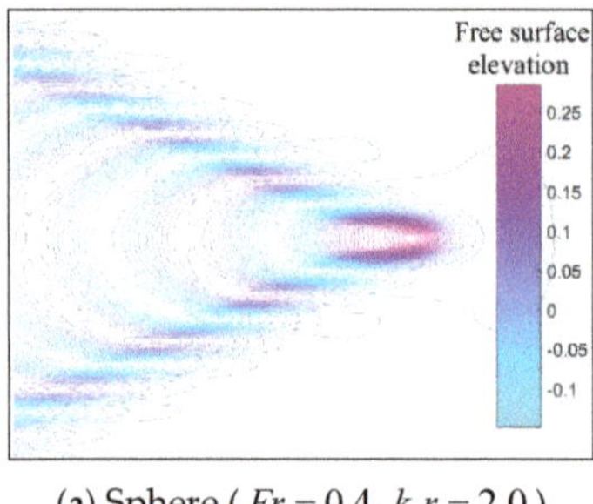

(a) Sphere ($Fr = 0.4$, $k_e r = 2.0$)

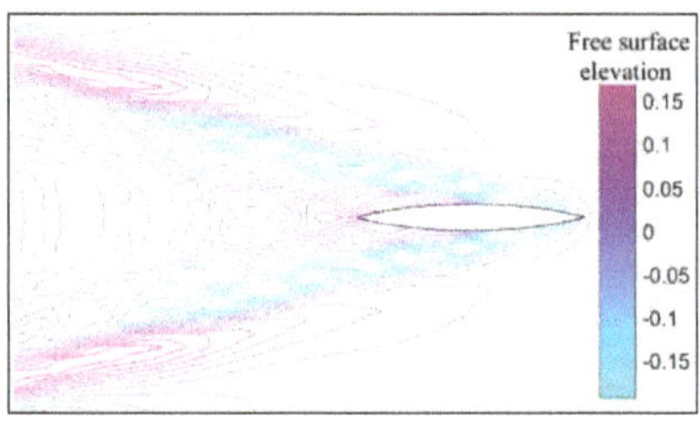

(b) Wigley I ship ($Fr = 0.4$, $\omega/\sqrt{g/L} = 3.0$)

Figure 9. Real part of the heave radiation wave contour of (**a**) submerged sphere and (**b**) Wigley I ship.

5. Conclusions

In this paper, a desingularized Rankine panel method based on the NSWM flow is applied for analysis of the hydrodynamic problems of a ship advancing in waves. NURBS are used to describe the body surface and the free surface. The wave exciting forces and the hydrodynamic coefficients are computed by solving the diffraction problem and radiation problem, respectively. A numerical study is carried out for a submerged sphere and a modified Wigley hull advancing in head waves. The numerical results are compared with the analytical solutions as well as other numerical results and experimental results available in literature. The following conclusions can be drawn.

(1) The numerical results of the wave exciting forces, added masses and damping coefficients computed using the present numerical method show good agreement with the published numerical and experimental results, which verifies the reliability of the present method. A comparison among the results indicates that the method based on the NSWM flow can generally give better agreement with the experimental and other published results than those based on NK and DB flows, especially for the hydrodynamic coefficients in relatively low frequency ranges.

(2) The NSWM potential has an influence on the prediction of the wave exciting forces. However, differences among different steady flow models are not very remarkable due to the dominant proportion of the Froude–Krylov force for the considered cases. The coupling effects between the nonlinear steady flow and the linear unsteady flow are important for the prediction of hydrodynamic coefficients, particularly at low frequencies.

(3) Compared with the time domain method, considering the NSWM flow as basic flow can be used as a more practical and faster numerical tool for evaluating the hydrodynamic performances of a ship in the early design stage.

In the present study, the method based on the NSWM flow is only applied for a submerged sphere and a modified Wigley hull. For reliable verification and application of this numerical method, further study on various ship forms needs to be carried out. Besides, in the numerical study, the squat of the hull (i.e., the trim and sinkage) is neglected. For the cases at larger forward speed, the numerical accuracy could be further improved by taking the effects of trim and sinkage into account. In addition, the desingularized Rankine panel method is only applied for the cases of Brard number τ larger than 0.25, where the radiation condition is satisfied by the staggered method. For $\tau < 0.25$, more robust methods for satisfying the radiation condition, such as the modified Sommerfeld radiation condition in [24,25], should be adopted. These will be the focuses of the future studies.

Author Contributions: Methodology, T.M.; formal analysis, T.M. and Z.Z.; Investigation, T.M.; writing—original draft preparation, T.M.; writing—review and editing, M.C., E.L. and Z.Z. All authors have read and agreed to the published version of the manuscript.

Funding: This research was funded by China Scholarships Council, grant number 201806230196; Lloyd's Register Foundation.

Acknowledgments: The first author gratefully acknowledges the financial support from China Scholarship Council (CSC), and from the Lloyd's Register Foundation (LRF) through the joint centre involving University College London, Shanghai Jiao Tong University, and Harbin Engineering University. LRF helps protect life and property by supporting engineering-related education, public engagement, and the application of research.

Conflicts of Interest: The authors declare no conflict of interest.

References

1. Ogilvie, T.F.; Tuck, E.O. *A Rational Strip Theory of Ship Motion, Part I*; Report No. 013; University of Michigan: Ann Arbor, MI, USA, 1969.
2. Salvesen, N.; Tuck, E.O.; Faltinsen, O.M. Ship motions and sea loads. *Trans. Soc. Nav. Archit. Mar. Eng.* **1970**, *78*, 250–287.
3. Kim, K.H.; Kim, Y.H. Comparative study on ship hydrodynamics based on Neumann-Kelvin and double-body linearizations in time-domain analysis. *Int. J. Offshore Polar Eng.* **2010**, *20*, 265–274.
4. Zhang, X.S.; Bandyk, P.; Beck, R.F. Seakeeping computations using double-body basis flows. *Appl. Ocean Res.* **2010**, *32*, 471–482. [CrossRef]
5. Zhang, X.S.; Beck, R.F. Fully nonlinear computations of wave radiation forces and hydrodynamic coefficients for a ship with a forward speed. In Proceedings of the 30th International Workshop on Water Waves and Floating Bodies, Bristol, UK, 12–15 April 2015.
6. Zhang, W.; Zou, Z.J. Time domain simulations of radiation and diffraction by a Rankine panel method. *J. Hydrodyn.* **2015**, *27*, 635–646. [CrossRef]
7. Seo, M.G.; Kim, Y.H. Numerical analysis on ship manoeuvring coupled with ship motion in waves. *Ocean Eng.* **2011**, *38*, 1934–1945. [CrossRef]
8. Zhang, W.; Zou, Z.J.; Deng, D.H. A study on prediction of ship manoeuvring in regular waves. *Ocean Eng.* **2017**, *137*, 367–381. [CrossRef]
9. Chen, X.; Zhu, R.C.; Zhao, J.; Zhou, W.J.; Fan, J. Study on weakly nonlinear motions of ship advancing in waves and influences of steady ship wave. *Ocean Eng.* **2018**, *150*, 243–257. [CrossRef]
10. Gao, Z.L.; Zou, Z.J. A NURBS-based high-order panel method for three-dimensional radiation and diffraction problems with forward speed. *Ocean Eng.* **2008**, *35*, 1271–1282. [CrossRef]
11. Bunnik, T.H.J. Seakeeping Calculations for Ships, Taking into Account the Non-linear Steady Waves. Ph.D. Thesis, Delft University of Technology, Delft, The Netherlands, 1999.
12. Söding, H.; Shigunov, V.; Schellin, T.E.; el Moctar, O. A Rankine panel method for added resistance of ships in waves. *J. Offshore Mech. Arct. Eng.* **2014**, *136*, 031601-1–031601-7. [CrossRef]
13. Chillcce, G.; el Moctar, O. A numerical method for manoeuvring simulation in regular waves. *Ocean Eng.* **2018**, *170*, 434–444. [CrossRef]
14. Riesner, M.; von Graefe, A.; Shigunov, V.; el Moctar, O. Prediction of non-linear ship responses in waves considering forward speed effects. *Ship Technol. Res.* **2016**, *63*, 135–145. [CrossRef]
15. Riesner, M.; el Moctar, O. A time domain boundary element method for wave added resistance of ships taking into account viscous effects. *Ocean Eng.* **2018**, *162*, 290–303. [CrossRef]
16. Cao, Y.; Schultz, W.; Beck, R. Three-dimensional desingularized boundary integral methods for potential problems. *Int. J. Numer. Methods Fluids* **1991**, *12*, 785–803. [CrossRef]
17. Beck, R.; Cao, Y.; Lee, T. Fully nonlinear water wave computations using the desingularized method. In Proceedings of the 6th International Conference on Numerical Ship Hydrodynamics, Iowa City, IA, USA, 2–5 August 1993.
18. Feng, A.C. A continuous desingularized source distribution method describing wave-body interactions of a large amplitude oscillatory body. *J. Offshore Mech. Arct. Eng.* **2015**, *137*, 021302-1–021302-10. [CrossRef]
19. Feng, A.C.; Bai, W.; You, Y.X.; Chen, Z.M.; Price, W.G. A Rankine source method solution of a finite depth, wave-body interaction problem. *J. Fluids Struct.* **2016**, *62*, 14–32. [CrossRef]
20. Feng, A.C.; Bai, W. Numerical simulation of wave radiation and diffraction problems with current effect. In Proceedings of the 12th International Offshore and Polar Engineering Conference, International Society of Offshore and Polar Engineers, Gold Coast, Australia, 4–7 October 2016.
21. Nakos, D.E.; Sclavounos, P.D. On steady and unsteady ship wave patterns. *J. Fluid Mech.* **1990**, *215*, 263–288. [CrossRef]

J. Mar. Sci. Eng. **2020**, *8*, 106

22. Jensen, G.; Mi, Z.X.; Söding, H. Rankine source methods for numerical solutions of steady wave resistance problem. In Proceedings of the 16th Symposium on Naval Hydrodynamics, Berkeley, CA, USA, 13–16 July 1986; pp. 575–582.
23. Aanesland, V. A hybrid model for calculating wave-making resistance. In Proceedings of the 5th International Conference on Numerical Ship Hydrodynamics, Hiroshima, Japan, 24–28 September 1989.
24. Das, S. Hydroelasticity of Marine Vessels Advancing in a Seaway. Ph.D. Thesis, University of Hawaii, Honolulu, HI, USA, 2011.
25. Das, S.; Cheung, K.F. Hydroelasticity of marine vessels advancing in a seaway. *J. Fluids Struct.* **2012**, *34*, 271–290. [CrossRef]
26. Journée, J.M.J. *Experiment and Calculations on Four Wigley Hullforms*; Report No. 909; Delft University of Technology, Ship Hydrodynamics Laboratory: Delft, The Netherlands, 1992.
27. Wu, G.X.; Taylor, R.E. Radiation and diffraction by a submerged sphere at forward speed. *Proc. R. Soc. Lond. A Math. Phys. Sci.* **1988**, *417*, 433–461.
28. Kim, W.D. Nonlinear free-Surface effects on a submerged sphere. *J. Hydronautics* **1969**, *3*, 29–37. [CrossRef]
29. Chen, X.B.; Malenica, S. Interaction effects of local steady flow on wave diffraction-radiation at low forward speed. *Int. J. Offshore Polar Eng.* **1998**, *8*, 102–109.
30. Kara, F.; Vassalos, D. Time domain prediction of steady and unsteady marine hydrodynamics problem. *Int. Shipbuild. Prog.* **2003**, *50*, 317–332.

Journal of
Marine Science and Engineering

Article

Development of a New Ship Adaptive Weather Routing Model Based on Seakeeping Analysis and Optimization

Silvia Pennino *, Salvatore Gaglione, Anna Innac, Vincenzo Piscopo and Antonio Scamardella

Department of Science and Technology, University of Naples "Parthenope", Centro Direzionale Isola C4, 80143 Naples, Italy; salvatore.gaglione@uniparthenope.it (S.G.); anna.innac@uniparthenope.it (A.I.); vincenzo.piscopo@uniparthenope.it (V.P.); antonio.scamardella@uniparthenope.it (A.S.)
* Correspondence: silvia.pennino@uniparthenope.it; Tel.: +39-081-547-6686

Received: 17 March 2020; Accepted: 8 April 2020; Published: 10 April 2020

Abstract: This paper provides a new adaptive weather routing model, based on the Dijkstra shortest path algorithm, aiming to select the optimal route that maximizes the ship performances in a seaway. The model is based on a set of ship motion-limiting criteria and on the weather forecast maps, providing the sea state conditions the ship is expected to encounter along the scheduled route. The new adaptive weather routing model is applied to optimize the scheduled route in the Northern Atlantic Ocean of the S175 containership, assumed as a reference vessel, based on the weather forecast data provided by the Global WAve Model (GWAM). In the analysis, both wave and combined wind/swell wave conditions are embodied to investigate the incidence on the optimum route assessment. Furthermore, the effect of the vessel speed on the optimum route detection is also investigated. Current results clearly show that it is possible to achieve appreciable improvements, up to 50% of the ship seakeeping performances, without excessively increasing the route length and the voyage duration.

Keywords: adaptive weather routing; Seakeeping Performance Index; route optimization; Dijkstra algorithm

---

1. Introduction

Maritime trades are strictly dependent on the weather conditions the ship is expected to encounter along her route, provided that excessive ship motions may cause damage or cargo loss, as well as the decrease of the onboard comfort level, with a negative impact on the safety of navigation for both cargo and passenger ships. In this respect, even if the route selection is entrusted to the ship master, the adaptive weather routing algorithms can provide a significant support to the decision-making process, to ensure the proper balancing between the safety of navigation and the related economic impact, in terms of voyage duration and fuel savings.

Nevertheless, as further discussed in Section 2, the impact of weather routing on fuel savings is expected to be marginal, provided that obtained results are not particularly encouraging, as no substantial fuel savings can be gathered if the ship route is varied, as regards the minimum distance route, without reducing the vessel speed. Hence, in the following research, a new adaptive weather routing model is developed to select the best route that maximizes the ship performances in a seaway, based on a set of proper seakeeping criteria. In fact, even if the route selection has a low impact in terms of fuel savings, an appreciable increase of the ship seakeeping performances can be gathered without excessively changing the route, as regards the minimum distance route. The new route optimization procedure, based on the shortest path algorithm, is developed in this paper to detect the best route, based on the weather forecasts the ship is expected to encounter up to the arrival in port. The algorithm is applied to the S715 containership, assumed as a reference vessel, to investigate the effect of the selected seakeeping parameters and vessel speed on the best route selection, as well as on the expected voyage duration and path length. Hence, the main aim of the current research is to develop a new adaptive weather routing model aiming to: (i) maximize the seakeeping performances in a seaway, (ii) investigate the incidence of unimodal/bimodal spectra and vessel speed on the improvement of the seakeeping performances, and (iii) carry out a sensitivity analysis to investigate the effect of each seakeeping parameter on the optimum route selection.

2. Literature Review

The weather routing problem was addressed by many authors and different approaches were embodied in the past. Following the pioneering works by James [1], Zoppoli [2], and Papadakis and Perakis [3], the first generation of weather routing criteria aimed to minimize the voyage duration, and consequently, the fuel consumption, mainly neglecting the impact of the best route selection on the ship performances in a seaway. Nevertheless, in the last years, the voyage optimization problem was approached in a wide sense, taking into account the ship seakeeping performances, apart from the fuel consumption. In this respect, the second-generation algorithms, developed as general optimization tools and applied to the ship voyage optimization problem, are generally categorized into: dynamic programming, genetic algorithms, and pathfinding methods.

The most used techniques include multi-objective genetic algorithms [4–7] that stochastically solve a discretized nonlinear optimization problem [8], based on the ship course and her velocity profile that, in turn, is represented by a set of parametric curves. In this respect, Szłapczynska and Smierzchalski [9] proposed a multi-criteria evolutionary weather-routing algorithm, based on an iterative process of population development, resulting in a Pareto-optimal set of solutions. Shao and Zhou [10] presented a three-dimensional (3D) dynamic-programming based on the original work by De Wit [11], devoted to optimizing the ship speed and the heading angle in the route planning. Zaccone et al. [12] developed a 3D dynamic-programming-based ship voyage optimization method, that attempts to select the optimal path and speed profile, based on the expected weather forecasting along the vessel route.

Another technique, devoted to managing the weather routing optimization procedure, is the path-finding method. In this respect, Padhy et al. [13] embodied the Dijkstra [14] algorithm to solve the ship routing problem, while Veneti [15] provided an improved solution based on the exact time-dependent bi-objective shortest path algorithm, to optimize two different conflicting objectives, namely, the fuel consumption and the expected risk along the route. In Reference [16], an overview of weather routing and safe ship handling approaches is presented, addressing the importance of the complementarity between both approaches in such situations and their extensive implementation to achieve optimal and safe navigation conditions in shipping. In Reference [17], some modifications are proposed to find the optimal path in ice-covered waters in order to take into account the possible involvement of an icebreaker. In this view, the icebreaker assistance is considered as an integral part of the overall formulation of the route optimization problem in the frame of graph-based and wave-based numerical routing algorithms.

A solution method for a real-life problem of routing and scheduling ships, motivated by the production and logistic activities of an oil company operating in Brazil, is described in Reference [18], where the authors proposed a multi-start heuristic that makes use of intensification and diversification strategies to minimize both transportation and docking costs, as well as avoid consecutive dockings of the same ship in different platforms. In Reference [19], the authors presented a study about a data-driven framework for decision support in the selection of optimal ship routes based on available weather forecasts and the calculation of fuel oil consumption over alternative routes. In detail, a modified version of the Dijkstra algorithm was recursively applied until the optimal route was detected.

3. Adaptive Weather Routing Model

3.1. Shortest Path Detection

The adaptive weather routing model, developed in Section 3, is based on the Dijkstra algorithm, that allows for the detection of the shortest path between two points of known coordinates, by maximizing a reference non-negative cost-function which is, in the current analysis, the Seakeeping Performance Index (*SPI*), provided by Equation (1). The Dijkstra algorithm is based on two problems, namely, the construction of the tree of minimum total length between a set of nodes, and the research of the best path between the start and the endpoint by minimizing an assigned cost function. The points, connected by a set of branches, form the required tree, from which a uniquely defined path between two nodes is obtained. In the current analysis, the cost function is the complement to 1 of the *SPI* provided by Equation (1). Besides, the algorithm is not applied statically but it is iteratively updated by changing the starting point, based on the elapsed time and the path already travelled by the ship. Before assessing the ship performances in a seaway, a network of nodes, between the start and the endpoints of the scheduled route, needs to be preliminarily provided to define a region that constrains the ship route. The network, generated as detailed by Lee et al. [20], is a special form of a Dijkstra graph consisting of nodes and arcs, associated to a set of parameters such as the path length, the cost, or the transit time. The network is generated according to the following steps: (i) the center point C of the great circle, black line in Figure 1a, between S and E, is determined, (ii) around C, a set of center points (red dots), with a fixed step distance, are considered, belonging to the rhumb-line perpendicular to the great circle, and (iii) a network of nodes, blue dots in Figure 1a, are generated, considering, in the first half of the great circle, a set of perpendicular rhumb-lines, as well as in the second half part. On each rhumb-line, a proper step distance between nodes is adopted. Each node is characterized by static information, such as the geographic coordinates, a dynamic grid with the weather data, namely, the significant wave height, the mean wave period, and the prevailing wave direction, which are systematically updated once a new dataset is available. After providing the network of nodes, a set of arcs, connecting two adjacent grid points, is created, and a cost-function, depending on the *SPI* values at the arc extremities, is assigned to detect the best route, as schematically shown in the flow chart reported in Figure 1b. Each segment of the optimal route corresponds to the maximum of the Seakeeping Performance Index among the set of possible values referenced to the calculation point.

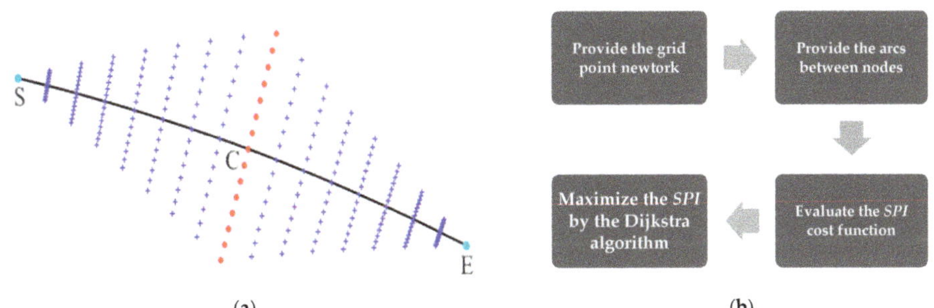

Figure 1. Adaptive weather routing algorithm: (**a**) Grid point network, (**b**) flow chart. *SPI*: Seakeeping Performance Index.

3.2. Assessment of Ship Seakeeping Performances

In the current analysis, the ship seakeeping performances are determined on the basis of five reference criteria, namely: (i) the amplitude of the pitch motion, (ii) the relative vertical acceleration at the ship forward perpendicular, (iii) the probability of slamming occurrence, (iv) the probability of green water on deck, and (v) the Motion Sickness Incidence (*MSI*), according to the following equation:

$$SPI = max\left\{0; \left(1 - \frac{rms_p}{rms_{p,l}}\right)\cdot\left(1 - \frac{rms_a}{rms_{a,l}}\right)\cdot\left(1 - \frac{p_{sl}}{p_{sl,l}}\right)\cdot\left(1 - \frac{p_{wd}}{p_{wd,l}}\right)\cdot\left(1 - \frac{MSI}{MSI_l}\right)\right\} \tag{1}$$

where rms_p (rms_a) is the RMS, root mean square, of the pitch motion amplitude (relative vertical acceleration), p_{sl} (p_{wd}) is the probability of occurrence of slamming (water on deck), *MSI* is the Motion Sickness Incidence, while the denominators represent the relevant limit values. It is noticed that when any seakeeping index is greater than the limit value, the *SPI* is null, so satisfying the non-negative condition is required to apply the Dijkstra method. The limit values of the pitch amplitude and the *MSI* are based on the NATO STANAG, Standardization Agreement, 4154 criteria, while the remaining ones comply with the NORDFORSK 1987 criteria [21,22], as detailed in Table 1.

Table 1. General operability limiting criteria for ships.

RMS of pitch amplitude	1.5 degrees
RMS of vertical acceleration at forward perpendicular	0.275 g (L ≤ 100 m) or 0.050 g (L ≥ 330 m) *
Slamming probability	0.03 (L ≤ 100 m) or 0.01 (L ≥ 330 m) *
Green water on deck probability	0.05
Motion Sickness Incidence	20% after 4 h

* at intermediate values, linear interpolation is applied.

The RMS of the pitch motion amplitude is determined according to the following equation, based on the ship Response Amplitude Operator [23]:

$$rms_p = \sqrt{\int_0^\infty \left|H_5(\omega_e)\right|^2 S_\varsigma(\omega_e)} \tag{2}$$

where H_5 is the speed-dependent pitch motion transfer function to be determined as detailed in Appendix A, S_ς is the wave spectrum, and $\omega_e = \omega - \omega^2\psi$ is the encounter wave frequency that satisfies the Doppler shift equation, depending on the absolute wave frequency ω and the factor $\psi = U\cos\mu/g$, where the vessel speed is denoted by U and μ denotes the heading angle. The RMS of the relative vertical acceleration is obtained by combining the heave and pitch motions at the ship forward perpendicular [23]:

$$rms_a = \sqrt{\int_0^\infty \left|H_3(\omega_e) - \bar{x}H_5(\omega_e) - e^{-ik\bar{x}\cos\mu}\right|^2 \omega_e^4 S_\varsigma(\omega_e)} \tag{3}$$

where H_3 is the heave motion transfer function, $\bar{x}$ is the longitudinal distance (fwd+) of the ship forward perpendicular from its center of mass, and $k = \omega^2/g$ is the wave number satisfying the deep-water condition. As concerns the slamming occurrence, it is determined according to the formula proposed by Faltinsen [24]:

$$p_{sl} = e^{-\left(\frac{v_{cr}^2}{2C_s^2 m_{2,r}} + \frac{d^2}{2C_s^2 m_{0,r}}\right)} \tag{4}$$

where $v_{cr} = 0.093\sqrt{gL}$ is the threshold velocity, C_s is the swell up coefficient equal to 1 up to $F_N = 0.30$ [25], and d is the ship draught at the forward perpendicular. In Equation (4), $m_{0,r}$ and $m_{2,r}$ denote the 0 and second-order spectral moments of the ship relative motion as regards the sea surface. As concerns the probability of green water on forward deck structures, it is determined in a quite similar manner [24]:

$$p_{wd} = e^{-\frac{f_b^2}{2C_s^2 m_{0,r}}} \tag{5}$$

where the freeboard at the ship forward perpendicular is denoted by f_b. Finally, the Motion Sickness Index is determined according to the formulation developed by O'Hanlon and McCauley [26] and modified by Colwell [27]:

$$MSI = 100\Phi(z_a)\Phi(z'_T) \tag{6}$$

where Φ is the standard normal cumulative distribution, while z_a and z'_T are determined by the following equations:

$$z_a = 2.128log_{10}\left(\frac{a_{MSI}}{g}\right) - 9.277log_{10}(f_m) - 5.809[log_{10}(f_m)]^2 - 1.851 \tag{7}$$

$$z'_T = 1.134z_a + 1.989log_{10}\left(\frac{T}{60}\right) - 2.904 \tag{8}$$

In Equations (7) and (8), T is the exposure time in s, while a_{MSI} and f_m are the RMS and the mean frequency in Hz of the ship vertical acceleration [27].

3.3. Assessment of Weather Forecasting Data

The availability of daily operational weather forecast data is a key point to assess the reference scenario required in the adaptive weather routing algorithm. In this respect, several software packages, with different resolutions, domains (from regional to global), and quality [28] are available, as proven by the agreements that shipping companies stipulate with one or more meteorological institutes, to obtain updated and reliable weather forecast data. Anyway, nowadays, the community generally follows the standardization established by the World Meteorological Organization (WMO), delivering all information in a self-descripting GRIB (GRIdded Binary) format [29], making the reading of the input data very easy. Among the variety of weather forecast codes, the third-generation Global WAve Model (GWAM), initially developed in the mid-80s by an international group of wave modelers [30], is probably the most embodied one by all research institutions around the world. It explicitly solves the wave transport equation, without any assumptions about the shape of the wave spectrum, therefore properly representing the physics of the wave evolution in compliance with the knowledge about the full set of freedom degrees of two-dimensional wave spectra. The model can be applied to any given regional or global grid, based on a certain topographic dataset. Besides, the grid resolution can be arbitrarily set in both space and time, while the wave propagation can be performed on latitudinal-longitudinal or Cartesian grids. The model outputs, embodied in the route optimization procedure, are the significant wave height and the mean wave period and direction of both wind wave and swell components.

4. Input Data

4.1. The S175 Containership

The S175 containership [31–33] is assumed as a reference vessel in the performed case study. The ship's main dimensions are listed in Table 2, while the zero-speed added mass and radiation damping are determined by the open source code Nemoh [34] and corrected to account for the forward speed effect, as detailed in Appendix A [35].

Table 2. Main particulars of the S175 containership.

Length between perpendiculars	175.0	m
Breadth	25.4	m
Design draught	9.5	m
Displacement	24,539	t
Pitch moment of inertia	43,286,796	tm^2
Waterplane area	3152	m^2
Longitudinal metacentric radius	206	m
Block coefficient	0.572	

4.2. Route Selection

The reference route falls in the North Atlantic Sea, as this trade lane is one of the most important routes in the global shipping industry and most of the key port infrastructures in Europe are continually upgraded to reflect the demands of shipping across this corridor. Besides, the weather conditions in this area can be rough and vary with seasons, influencing the obtained results. The selected voyage has (34° N, 60° W) departure and (32° N, 20° W) arrival coordinates, as depicted in Figure 2, that also provides the great circle or orthodrome (black) and the rhumb-line or loxodrome (red) routes, equal to 2005 and 2018 nm respectively, provided that they are generally combined by the master during the navigation [36,37]. In fact, the great circle or orthodrome is the shortest path route, but it requires to constantly change the vessel heading. On the contrary, the rhumb-line or loxodrome is slightly longer, but it allows the master to keep a constant heading in the absence of wind and sea current, as they are generally combined for practical purposes. The voyage duration is equal to about 7 days, considering a reference vessel speed of 12 knots, corresponding to Froude number $F_N = 0.15$. From Figure 2, it is gathered that the start and the endpoint are slightly far from the departure and arrival ports. This choice is mainly due to the marine traffic congestion the ship experiences when it approaches the coastline, that makes the application of adaptive routing models challenging, as additional restraints, mainly related to neighboring vessels, arise.

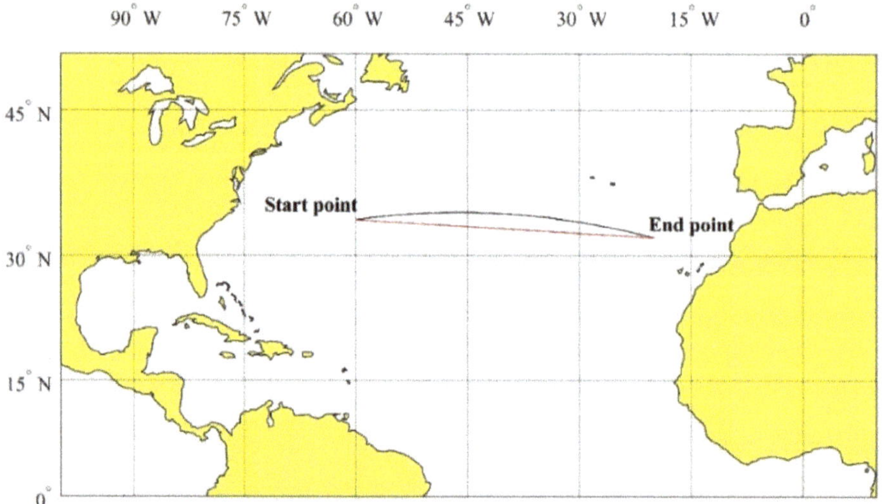

Figure 2. Great circle (black) and rhumb-line (red) between routes' departure and arrival points.

4.3. Weather Forecasting Data

In the current analysis, the GRIB files were downloaded for the period from 3 up to 10 January 2020, with 0.25° × 0.25° grid spacing, 3-hour forecast interval, and two observation periods equal to 1 and 7 days, respectively. The free software XyGrib, able to download and show the meteorological data, was embodied to obtain the GRIB files. An example of the images providing the changes of the environmental conditions in the selected area is reported in Figure 3, that provides a series of GRIB files showing the time trend of the significant wave height, $H_s = \sqrt{H_{s,wind}^2 + H_{s,swell}^2}$ from 06:00 UTC (Coordinated Universal Time) on 3 January 2020 up to 03:00 UTC on 4 January 2020, where the wind wave (swell) component is denoted by $H_{s,wind}$ ($H_{s,swell}$).

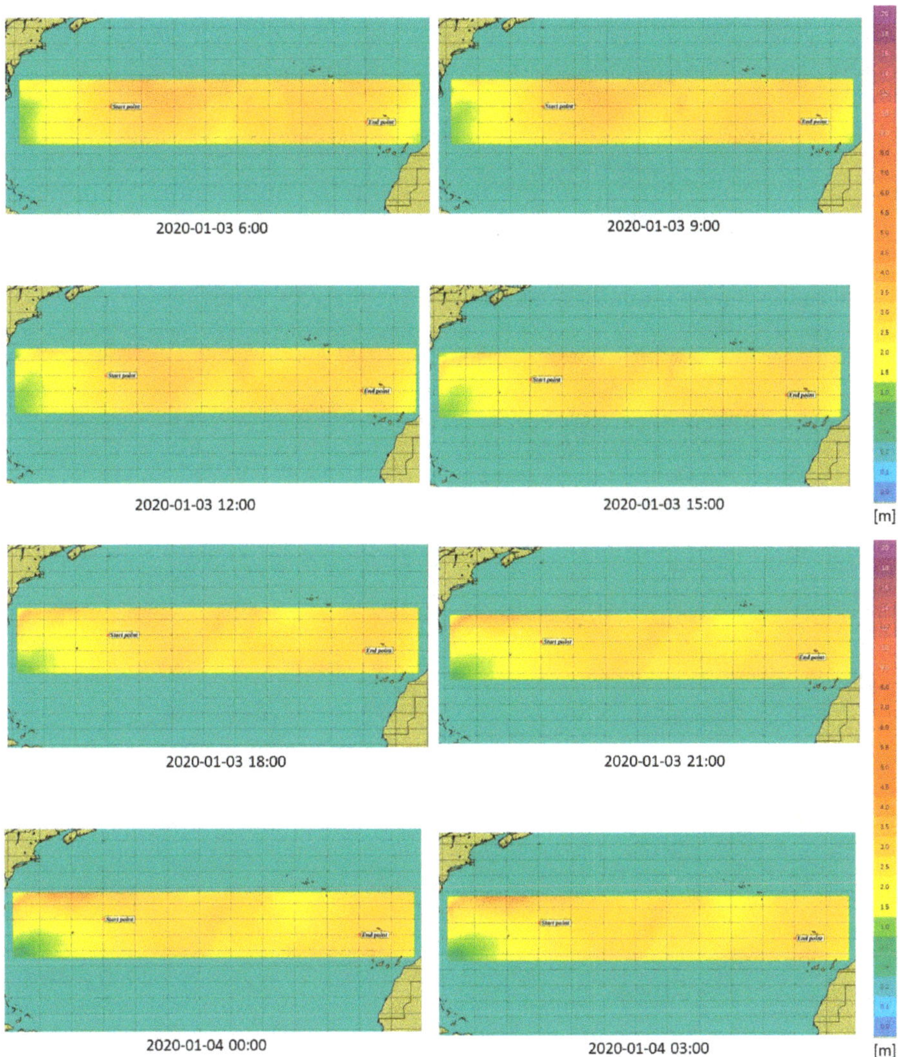

Figure 3. Time trend of the significant wave height from 06:00 UTC on 3 January 2020 up to 03:00 UTC on 4 January 2020.

5. Case Study

5.1. Optimum Route Detection

After defining the input data, two GRIB files are embodied in the route optimization procedure. The former, referenced as "Case 1", is a 7-day forecast GRIB file, ranging from 3 up to 10 January 2020. The latter, referenced as "Case 2", is a 1-day GRIB file, updated daily in the same reference period, according to the effective position of the ship during the voyage. The same GRIB files, referenced as "Case 3" and "Case 4", are re-analyzed, including the swell data, apart from the wind-generated waves. In the first two cases, the JONSWAP [38] spectrum is applied in the weather routing optimization procedure, while in the remaining ones, the two-peak Torsethaugen [39] spectrum is embodied, to model weather conditions obtained by combined wind sea and swell components, coming from different directions.

In this respect, Figure 4a–d provide the optimal routes with reference to the four selected conditions. In all cases, the great circle route is highlighted in red, the optimum route is depicted in black, while the grey circles represent the starting and the ending points. Besides, Table 3 provides the difference, Δ_c, between the great circle and optimum route, that maximizes the ship performances, together with the percentage variation, Δ_{SPI}, of the Seakeeping Performance Index (SPI), as regards the values corresponding to the great circle route. Based on the current results, the percentage increase of the route length ranges from 1.1% up to 2.7%, while the SPI increase is much higher and ranges from 10% up to 40%. These results clearly show that the ship seakeeping performances can be highly improved, without significantly increasing the voyage length and, consequently, the fuel consumptions.

After carrying out this preliminary analysis, Figure 5a–d and Table 4 provide the same calculations obtained by reversing the travel direction, therefore inverting the departure and arrival coordinates. Also, in this case, the percentage increase of the route length is low and ranges from 3.1% up to 4.9%, while the SPI increase is much higher and ranges from about 10% up to 68%. All calculations were performed based on a reference vessel speed equal to 12 kn, corresponding to $F_N = 0.15$.

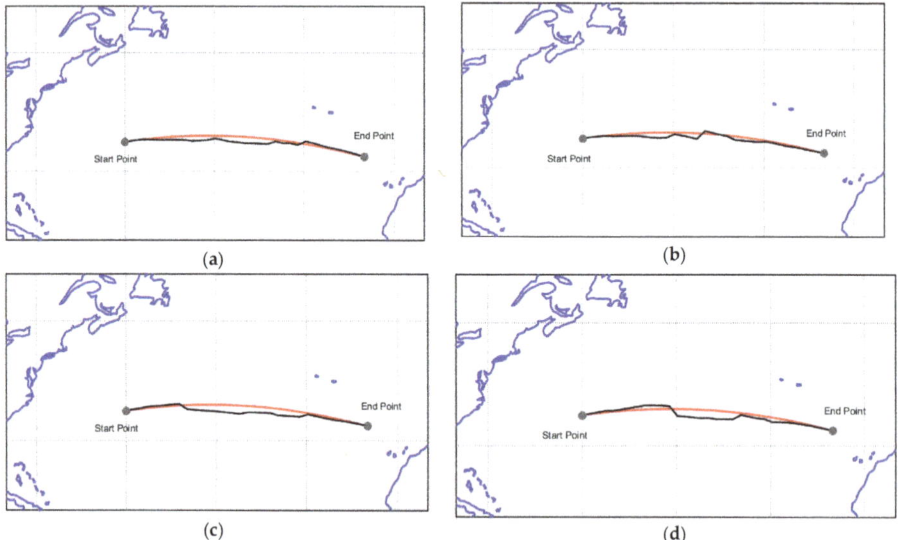

Figure 4. Minimum distance (great circle) and optimal route detection, $F_N = 0.15$: (**a**) Case 1; (**b**) Case 2; (**c**) Case 3; (**d**) Case 4.

Table 3. Results of optimal route detection, $F_N = 0.15$.

Case	GRIB File Update	Spectrum	Δ_c nm	Δ_c %	Δ_{SPI} %
1	7-day	JONSWAP	22.5	1.1	40.5
2	1-day	JONSWAP	47.3	2.4	26.6
3	7-day	Torsethaugen	26.6	1.3	6.6
4	1-day	Torsethaugen	53.6	2.7	9.4

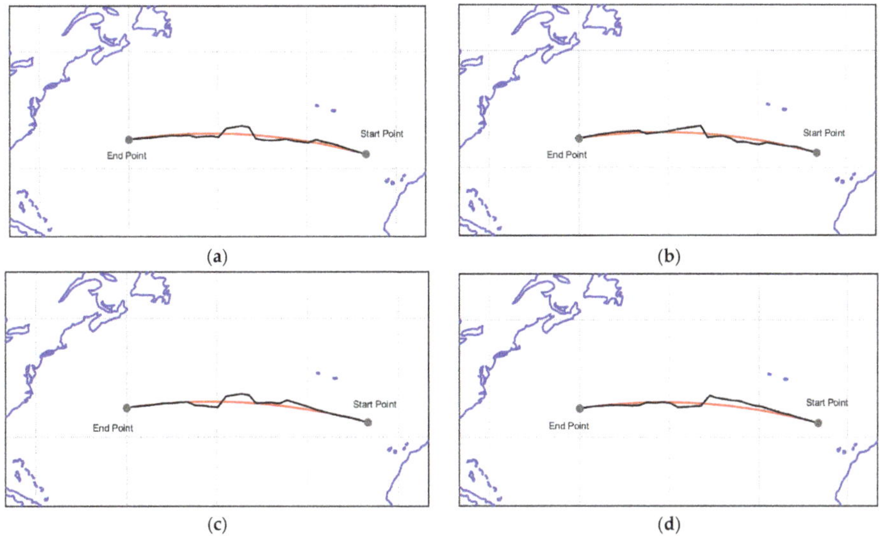

Figure 5. Minimum distance (great circle) and optimal route detection, $F_N = 0.15$ (reversed travel direction): (**a**) Case 1; (**b**) Case 2; (**c**) Case 3; (**d**) Case 4.

Table 4. Results of optimal route detection, $F_N = 0.15$ (reversed travel direction).

Case	GRIB File Update	Spectrum	Δ_c nm	Δ_c %	Δ_{SPI} %
1	7-day	JONSWAP	98.9	4.9	22.7
2	1-day	JONSWAP	72.9	3.6	51.1
3	7-day	Torsethaugen	91.1	4.5	32.5
4	1-day	Torsethaugen	63.0	3.1	68.3

5.2. Effect of Vessel Speed

The incidence of the vessel speed on the optimum route detection is investigated, increasing the Froude number from 0.18 up to 0.24, with 0.03 step. Obviously, the vessel speed increase leads to a corresponding decrease of the travelling time that diminishes up to 138, 119, and 104 h, as well as to a slight variation of the weather conditions that the ship is expected to encounter during the voyage. In this respect, Figure 6a–d, Figures 7a–d and 8a–d provide the optimal routes, while Table 5 summarizes the obtained results. In the first case, corresponding to $F_N = 0.18$, the route length increases from 0.6% up to 2.0%, if compared with the great circle one, while the *SPI* increases from 3.5% up to 31.5%. In the second condition, corresponding to $F_N = 0.21$, the route length rises up from 0.6% to 6.1%, while the *SPI* increase ranges from 3.4% up to 39.8%. In the last condition, instead, the route length increases from 0.3% up to 1.6% and the *SPI* rises from 6.7% up to 32.1%. Based on the current results, it is confirmed that it is possible to achieve considerable improvements of the ship seakeeping performances, without significantly affecting the voyage length. Furthermore, it is verified that the vessel speed plays a fundamental role in the assessment of the optimal route, as well as on the increase of the ship seakeeping performances.

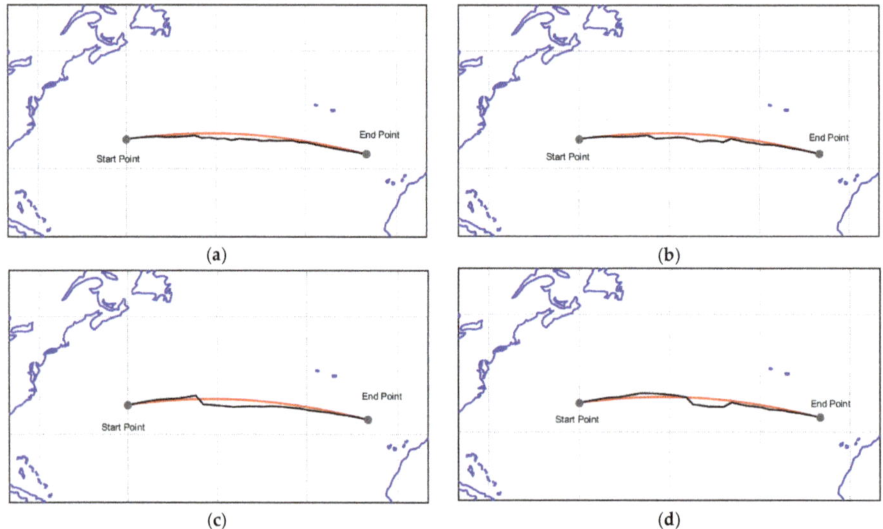

Figure 6. Minimum distance (great circle) and optimal route detection, $F_N = 0.18$: (**a**) Case 1; (**b**) Case 2; (**c**) Case 3; (**d**) Case 4.

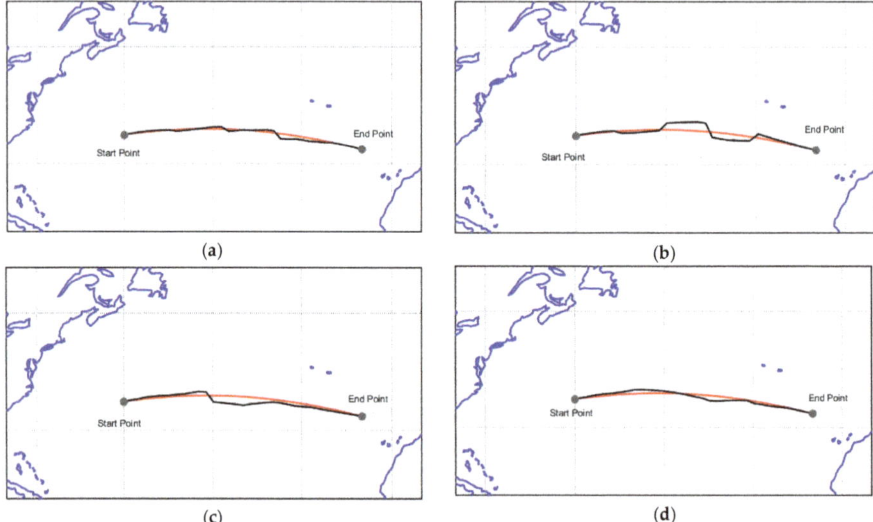

Figure 7. Minimum distance (great circle) and optimal route detection, $F_N = 0.21$: (**a**) Case 1; (**b**) Case 2; (**c**) Case 3; (**d**) Case 4.

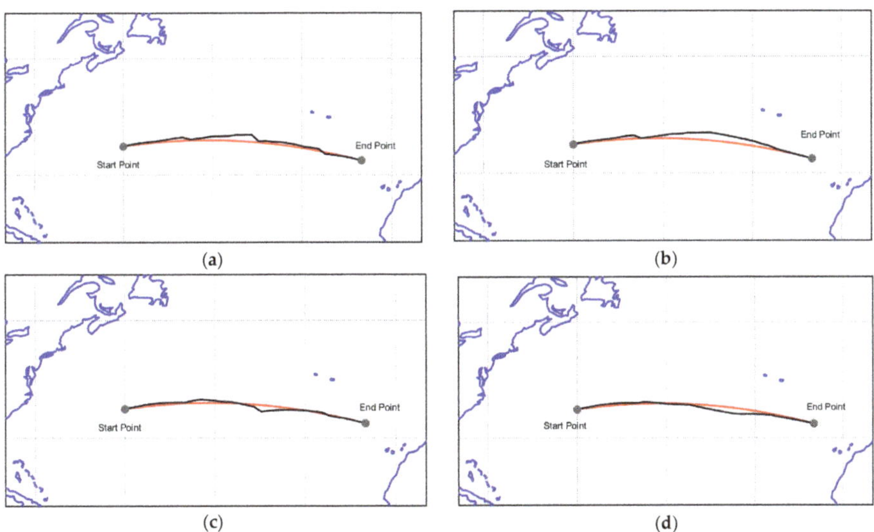

(a) (b) (c) (d)

Figure 8. Minimum distance (great circle) and optimal route detection, $F_N = 0.24$: (**a**) Case 1; (**b**) Case 2; (**c**) Case 3; (**d**) Case 4.

Table 5. Incidence of vessel speed on optimal route detection.

	Reference Conditions		$F_N = 0.18$			$F_N = 0.21$			$F_N = 0.24$		
Case	File Update	Spectrum	Δ_c nm	Δ_c %	Δ_{SPI} %	Δ_c nm	Δ_c %	Δ_{SPI} %	Δ_c nm	Δ_c %	Δ_{SPI} %
1	7-day	JONSWAP	12.1	0.6	22.4	38.2	1.9	33.1	32.1	1.6	51.5
2	1-day	JONSWAP	21.2	1.1	31.5	122.7	6.1	39.8	13.0	0.6	36.1
3	7-day	Torsethaugen	37.7	1.9	3.5	47.3	2.4	3.4	19.3	1.0	2.4
4	1-day	Torsethaugen	40.0	2.0	11.5	11.8	0.6	19.2	32.1	1.6	51.5

5.3. Sensitivity Analysis

The incidence on the optimum route detection of each seakeeping index provided by Equation (1) is investigated. The optimum routes, corresponding to $F_N = 0.15$, are reported in Figure 9a–j with reference to both JONSWAP (left) and Torsethaugen (right) wave spectra, separately considering the following criteria, namely: (i) the RMS of pitch amplitude (a,b), (ii) the RMS of vertical acceleration (c,d), (iii) the slamming probability (e,f), (iv) the water on deck probability (g,h), and (v) the *MSI* (i,j).

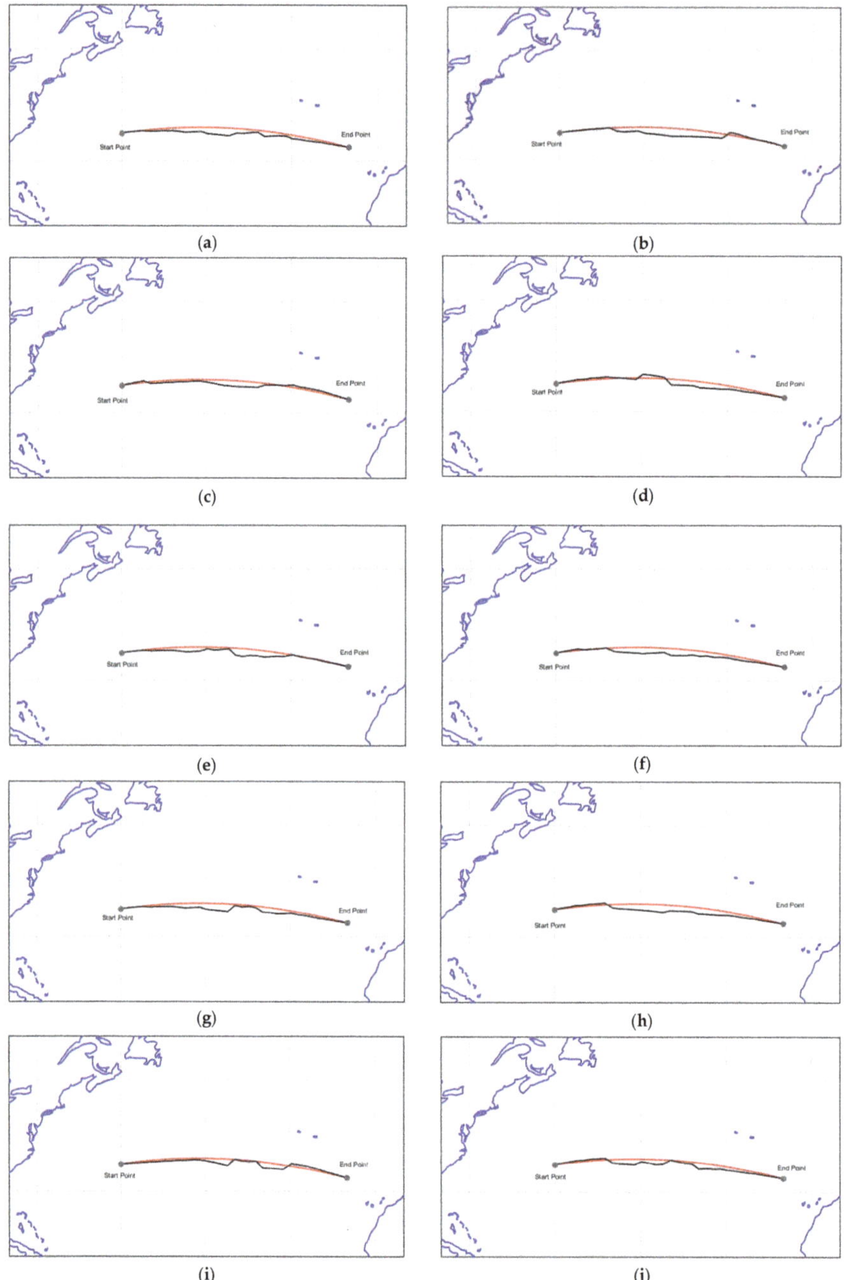

Figure 9. Sensitivity analysis at $F_N = 0.15$ based on JONSWAP (left) and Torsethaugen (right) wave spectra: (**a**) RMS of pitch amplitude—JONSWAP; (**b**) RMS of pitch amplitude—Torsethaugen; (**c**) RMS of vertical acceleration—JONSWAP; (**d**) RMS of vertical acceleration—Torsethaugen; (**e**) Slamming probability—JONSWAP; (**f**) Slamming probability—Torsethaugen; (**g**) Water on deck probability—JONSWAP; (**h**) Water on deck probability—Torsethaugen; (**i**) *MSI*—JONSWAP; (**j**) *MSI*—Torsethaugen.

Table 6 provides a comparative analysis between the great circle and the optimum route, together with the percentage variation of the *SPI*. Based on the current results, the RMS of the vertical acceleration and the Motion Sickness Incidence seem to be the most sensitive parameters. This outcome is confirmed with reference to both JONSWAP and Torsethaugen spectra, even if in the latter case the improvements of the seakeeping performances are slightly lower.

Table 6. Sensitivity analysis, 7-day, $F_N = 0.15$.

Seakeeping Parameter	JONSWAP			Torsethaugen		
	Δ_c (nm)	Δ_c (%)	Δ_{SPI} (%)	Δ_c (nm)	Δ_c (%)	Δ_{SPI} (%)
RMS of pitch amplitude	23.2	1.2	2.2	37.0	1.8	1.4
RMS of vertical acceleration	17.6	0.9	11.0	44.6	2.2	12.2
Slamming probability	55.3	2.8	1.5	18.8	0.9	1.8
Water on deck probability	53.6	2.7	1.8	21.5	1.1	8.3
Motion Sickness Incidence	72.0	3.6	29.7	47.1	2.3	10.1

6. Discussion

The results obtained in Section 5, obtained by a purposely developed code in Matlab, MathWorks, clearly highlight that there is wide space to improve the ship performances in a seaway by properly varying the vessel route, depending on the expected met-ocean conditions. In this respect, the following main outcomes have been achieved:

(i)   The selection of the seakeeping indexes represents a basic point in the development and application of the adaptive weather routing model, as well as in the selection of the optimum route that maximizes the *SPI* provided by Equation (1).

(ii) The employment of 7-day and 1-day GRIB files plays a fundamental role in the assessment of the optimum route, which implies that the update frequency of the met-ocean conditions is a key point and it should be as high as possible.

(iii) The vessel speed and the combined presence of wind wave and swell components affect the selection of the optimum route and the improvement of the seakeeping performances in a seaway.

(iv) The sensitivity analysis highlights that the optimum route significantly varies if the seakeeping parameters are embodied separately in the adaptive weather routing model.

These outcomes clearly suggest that the adaptive weather routing model shall be specialized on a case-by-case basis, depending on the ship type and the key seakeeping performances that need to be monitored and improved.

7. Conclusions

The Dijkstra shortest path algorithm was applied to detect the optimum route, constrained to maximize a non-negative cost function, namely the *SPI* index provided by Equation (1), in order to maximize the ship performances in a seaway. Based on the current results, the proposed adaptive weather routing algorithm allows for consistently increasing the ship seakeeping performances up to 50%, without significantly affecting the voyage duration or the route length. Furthermore, it was verified that the vessel speed plays a fundamental role in the assessment of the optimum route, as well as on the improvement of the ship seakeeping performances. Besides, the existence of a swell component, coming from a different direction as regards the wind wave one, can affect the selection of the optimal route. Following these outcomes, the application of an adaptive weather routing criterion, based on ship seakeeping analysis and optimization, seems to be the most suitable way to detect the optimal route, provided that the currently embodied models have an almost negligible impact in terms of fuel savings. Nevertheless, additional fuel savings, also resulting in low gas emissions in the atmosphere, could be gathered by properly varying the ship speed, depending on the estimated time of arrival in port and the availability of the harbor infrastructures. In this respect, if the ship has to wait at anchor, due to the unavailability of the harbor infrastructures for the scheduled loading/unloading operations, it is certainly preferable to reduce the vessel speed, resulting in fuel savings, and delay the time of arrival in port. All these suggestions could be embodied by ship-owners and port authorities, in order to establish a cooperating network between the harbor and the onboard vessel management, resulting in: (i) an increase of the ship performances in a seaway with a positive impact on the safety of navigation, (ii) a reduction of the fuel consumptions, resulting in savings for the ship-owner and in a possible reduction of the gas emissions in the atmosphere, and (iii) optimization of the scheduled loading/unloading operations in port.

Based on these outcomes, the current results are encouraging for further research activities, devoted to including additional seakeeping criteria in the adaptive weather routing algorithm. In this respect, the incidence of the selected seakeeping criteria on the optimum route assessment needs to be further investigated. In fact, the *SPI* index mainly depends on the ship type and mission, which implies that the proposed adaptive weather routing model needs to be properly specialized, on a case-by-case basis. Finally, the current results were obtained in a simulated environment, and therefore, they need to be further verified in real conditions, in order to carry out a comparative analysis between the optimum route, based on the current seakeeping adaptive weather routing model, and previous real ship routes. These topics will be the subject of future works.

Author Contributions: Conceptualization, S.G., V.P., and A.S.; Writing—original draft, S.P. and A.I.; Writing—review and editing, S.G., V.P., and A.S.; Methodology, S.P. and A.I.; Software and formal analysis, S.P.; Data curation and visualization, A.I.; Supervision, S.G., V.P., and A.S.; Validation S.G. and V.P.; Funding acquisition, A.S. All authors have read and agreed to the published version of the manuscript.

Funding: This research received no external funding.

Acknowledgments: The work was supported by "DORA—Deployable Optics for Remote Sensing Applications" (ARS01_00653), a project funded by MIUR-PON "Research & Innovation"/PNR 2015–2020.

Conflicts of Interest: The authors declare no conflict of interest.

Appendix A

The heave/pitch motion equations in the encounter wave frequency-domain are provided by the following set of equations:

$$\begin{cases} \left[-\omega_e^2(\Delta + A_{33}) + i\omega_e B_{33} + C_{33}\right]H_3 + \left[-\omega_e^2 A_{35} + i\omega_e B_{35} + C_{35}\right]H_5 = f_3(\omega) \\ \left[-\omega_e^2 A_{53} + i\omega_e B_{53} + C_{53}\right]H_3 + \left[-\omega_e^2(I_{55} + A_{55}) + i\omega_e B_{55} + C_{55}\right]H_5 = f_5(\omega) \end{cases} \tag{A1}$$

where Δ and I_{55} are the ship displacement and pitch moment of inertia, A_{ij}, B_{ij}, and C_{ij} are the added mass, hydrodynamic damping, and restoring coefficient for the i–jth mode, H_3 and H_5 are the heave and pitch motion transfer functions, while f_3 and f_5 are the heave force and pitch moment per unit wave amplitude. After solving the Equation System (A1), the modulus of the heave and pitch motion transfer functions is derived [23]:

$$|H_3(\omega_e)| = \left| \frac{\left[-\omega_e^2(I_{55}+A_{55})+i\omega_e B_{55}+C_{55}\right]f_3(\omega)-\left[-\omega_e^2 A_{35}+i\omega_e B_{35}+C_{35}\right]f_5(\omega)}{\left[-\omega_e^2(\Delta+A_{33})+i\omega_e B_{33}+C_{33}\right]\left[-\omega_e^2(I_{55}+A_{55})+i\omega_e B_{55}+C_{55}\right]-\left[-\omega_e^2 A_{35}+i\omega_e B_{35}+C_{35}\right]\left[-\omega_e^2 A_{53}+i\omega_e B_{53}+C_{53}\right]} \right| \tag{A2}$$

$$|H_3(\omega_e)| = \left| \frac{\left[-\omega_e^2(\Delta+A_{33})+i\omega_e B_{33}+C_{33}\right]f_5(\omega)-\left[-\omega_e^2 A_{53}+i\omega_e B_{53}+C_{53}\right]f_3(\omega)}{\left[-\omega_e^2(\Delta+A_{33})+i\omega_e B_{33}+C_{33}\right]\left[-\omega_e^2(I_{55}+A_{55})+i\omega_e B_{55}+C_{55}\right]-\left[-\omega_e^2 A_{35}+i\omega_e B_{35}+C_{35}\right]\left[-\omega_e^2 A_{53}+i\omega_e B_{53}+C_{53}\right]} \right| \tag{A3}$$

In Equations (A2) and (A3), the frequency-dependent added masses and radiation dampings combine the zero-speed values with the speed-dependent corrective factors, with no transom correction [35]:

$$A_{33} = A_{33}^0(\omega_e); \ B_{33} = B_{33}^0(\omega_e)$$
$$A_{35} = A_{35}^0(\omega_e) - \frac{U}{\omega_e^2}B_{33}^0(\omega_e); \ B_{35} = B_{35}^0(\omega_e) + UA_{33}^0(\omega_e)$$
$$A_{53} = A_{53}^0(\omega_e) + \frac{U}{\omega_e^2}B_{33}^0(\omega_e); \ B_{53} = B_{53}^0(\omega_e) - UA_{33}^0(\omega_e) \tag{A4}$$
$$A_{55} = A_{55}^0(\omega_e) + \frac{U^2}{\omega_e^2}A_{33}^0(\omega_e); \ B_{55} = B_{55}^0(\omega_e) + \frac{U^2}{\omega_e^2}B_{33}^0(\omega_e)$$

If the 1-to-3 multivalued problem between absolute and encounter wave frequencies occurs, the modulus of heave/pitch motion transfer functions is replaced by the following equation:

$$\left|H_j(\omega_e)\right| = \sqrt{\left|H_j^1(\omega_e)\right|^2 + \left|H_j^2(\omega_e)\right|^2 + \left|H_j^3(\omega_e)\right|^2} \tag{A5}$$

where $H_j^i(\omega_e)$ is the j-th mode transfer function at the absolute wave frequency ω_i, with $i = 1, 2,$ or 3.

References

1. James, R.W. *Application of Wave Forecasts to Marine Navigation*; U.S. Naval Oceanographic Office: Washington, DC, USA, 1957.
2. Zoppoli, R. Minimum-Time Routing as an n-Stage Decision Process. *J. Appl. Meteorol.* **1972**, *11*, 429–435. [CrossRef]
3. Papadakis, N.A.; Perakis, A.N. Deterministic Minimal Time Vessel Routing. *Oper. Res.* **1990**, *38*, 426–438. [CrossRef]
4. Marie, S.; Courteille, E. Multi-objective optimization of motor vessel route. *Saf. Sea Transp.* **2009**, *3*, 133–141.
5. Maki, A.; Akimoto, Y.; Nagata, Y.; Kobayashi, S.; Kobayashi, E.; Shiotani, S.; Ohsawa, T.; Umeda, N. A new weather-routing system that accounts for ship stability based on a real-coded genetic algorithm. *J. Mar. Sci. Technol.* **2011**, *16*, 311–322. [CrossRef]
6. Vettor, R.; Soares, C.G. Development of a ship weather routing system. *Ocean Eng.* **2016**, *123*, 1–14. [CrossRef]
7. Zaccone, R.; Figari, M.; Altosole, M.; Ottaviani, E.; Soares, C.; Santos, T. Fuel saving-oriented 3D dynamic programming for weather routing applications. In Proceedings of the Maritime Technology and Engineering III, MARTECH 2016, Lisbon, Portugal, 4–6 July 2016; pp. 183–189.
8. Hinnenthal, J. Robust Pareto Optimum Routing of Ships Utilizing Deterministic and Ensemble Weather Forecasts. Ph.D. Thesis, Technischen Universität Berlin, Berlin, Germany, 2008.
9. Szłapczyńska, J.; Smierzchalski, R. Multicriteria optimisation in weather routing. *Saf. Sea Transp.* **2009**, *3*, 393.
10. Weintrit, A.; Neumann, T.; Wei, S.; Zhou, P. Development of a 3D Dynamic Programming Method for Weather Routing. *Methods Algorithms Navig.* **2011**, *6*, 181–187.
11. De Wit, C. Proposal for Low Cost Ocean Weather Routeing. *J. Navig.* **1990**, *43*, 428. [CrossRef]
12. Zaccone, R.; Ottaviani, E.; Figari, M.; Altosole, M. Ship voyage optimization for safe and energy-efficient navigation: A dynamic programming approach. *Ocean Eng.* **2018**, *153*, 215–224. [CrossRef]
13. Padhy, C.P.; Sen, D.; Bhaskaran, P.K. Application of wave model for weather routing of ships in the North Indian Ocean. *Nat. Hazards* **2007**, *44*, 373–385. [CrossRef]
14. Dijkstra, E.W. A note on two problems in connexion with graphs. *Numer. Math.* **1959**, *1*, 269–271. [CrossRef]
15. Veneti, A.; Makrygiorgos, A.; Konstantopoulos, C.; Pantziou, G.; Vetsikas, I.A. Minimizing the fuel consumption and the risk in maritime transportation: A bi-objective weather routing approach. *Comput. Oper. Res.* **2017**, *88*, 220–236. [CrossRef]
16. Perera, L.P.; Soares, C.G. Weather routing and safe ship handling in the future of shipping. *Ocean Eng.* **2017**, *130*, 684–695. [CrossRef]
17. Topaj, A.; Tarovik, O.; Bakharev, A.; Kondratenko, A. Optimal ice routing of a ship with icebreaker assistance. *Appl. Ocean Res.* **2019**, *86*, 177–187. [CrossRef]
18. Yamashita, D.; Da Silva, B.J.V.; Morabito, R.; Ribas, P.C. A multi-start heuristic for the ship routing and scheduling of an oil company. *Comput. Ind. Eng.* **2019**, *136*, 464–476. [CrossRef]
19. Gkerekos, C.; Lazakis, I. A novel, data-driven heuristic framework for vessel weather routing. *Ocean Eng.* **2020**, *197*, 106887. [CrossRef]

20. Lee, H.; Kong, G.; Kim, S.; Kim, C.; Lee, J. Optimum Ship Routing and It's Implementation on the Web. In *Computer Vision*; Springer Science and Business Media LLC: Berlin, Germany, 2002; Volume 2402, pp. 125–136.

21. Stevens, S.C.; Parsons, M.G. Effects of Motion at Sea on Crew Performance: A Survey. *Mar. Technol.* **2002**, *39*, 29–47.

22. Pipchenko, O.D.; Zhukov, D.S. Ship Control Optimization in Heavy Weather Conditions. In Proceedings of the 2010 11th AGA, IAMU, Busan, Korea, 16–18 October 2010; pp. 91–95.

23. Lewis, E.V. *Principles of Naval Architecture 2nd Revision Vol. III*; The Society of Naval Architects and Marine Engineers: Jersey City, NJ, USA, 1989.

24. Faltinsen, O.M. *Sea Loads on Ships and Offshore Structures*; Cambridge University Press: Cambridge, UK, 1990.

25. Blok, J.J.; Huisman, J. Relative motions and swell-up for a frigate bow. *R. Inst. Nav. Arch. Trans.* **1984**, *126*, 227–244.

26. O'Hanlon, J.F.; McCauley, M.E. Motion sickness as a function of the frequency and acceleration of vertical of vertical sinusoidal motion. *Aerosp. Med.* **1974**, *45*, 366–369.

27. Colwell, J.L. *Human Factors in the Naval Environment: A Review of Motion Sickness and Biodynamic Problems*; National Defence Technical Memorandum 89/220; Defence Centre de Research Establishment Atlantic: Dartmouth, NS, Canada, 1989.

28. Bidlot, J.; Holmes, D.J.; Wittmann, P.A.; Lalbeharry, R.; Chen, H.S. Intercomparison of the Performance of Operational Ocean Wave Forecasting Systems with Buoy Data. *Weather. Forecast.* **2002**, *17*, 287–310. [CrossRef]

29. WMO. Introduction to GRIB Edition 1 and GRIB Edition 2. Available online: https://www.wmo.int/pages/prog/www/WMOCodes/Guides/GRIB/Introduction_GRIB1-GRIB2.pdf (accessed on 18 November 2019).

30. Komen, G.J.; Cavaleri, L.; Donelan, M.; Hasselmann, K.; Janssen, P.A.E.M.; Hasselmann, S. *Dynamics and Modelling of Ocean Waves*; Cambridge University Press (CUP): Cambridge, UK, 1994.

31. ITTC. Report of the Seakeeping Committee. In Proceedings of the 15th International Towing Tank Conference, The Hague, The Netherlands, 11–15 September 2006; Volume I, pp. 55–114.

32. Fonseca, N.; Guedes Soares, C. Time-domain analysis of large amplitude vertical ship motions and wave loads. *J. Ship Res.* **1998**, *42*, 139–153.

33. Kim, M.; Hizir, O.; Turan, O.; Day, A.; Incecik, A. Estimation of added resistance and ship speed loss in a seaway. *Ocean Eng.* **2017**, *141*, 465–476. [CrossRef]

34. Babarit, A.; Delhommeau, G. Theoretical and numerical aspects of the open source BEM solver NEMOH. In Proceedings of the 11th European Wave and Tidal Energy Conference, Nantes, France, 6–11 September 2015.

35. Salvesen, N.; Tuck, E.O.; Faltinsen, O. Ship motions and sea loads. *SNAME Trans.* **1970**, *6*, 1–30.

36. Bowditch, N. *The American Practical Navigator—An Epitome of Navigation*; National Imagery and Mapping Agency: Bethesda, MD, USA, 2002.

37. Eskild, H. Development of a Method for Weather Routing of Ships. Master's Thesis, NTNU Institutt for Marin Teknikk, Trondheim, Norway, 2014.

38. Hasselmann, K.; Barnett, T.P.; Bouws, E.; Carlson, H.; Cartwright, D.E.; Enke, K.; Ewing, J.A.; Gienapp, H.; Hasselmann, D.E.; Kruseman, P.; et al. *Measurement of Wind Wave Growth and Swell Decay during the Joint North Sea Wave Project (JONSWAP)*; Deutches Hydrographisches Institut: Hamburg, Germany, 1973.

39. Torsethaugen, K. A two-peak wave spectral model. In Proceedings of the 1993 12th International Conference on Offshore Mechanics and Arctic Engineering, Glasgow, UK, 20–24 June 1993; Volume 2, pp. 175–180.

Article

Vertical Motions Prediction in Irregular Waves Using a Time Domain Approach for Hard Chine Displacement Hull

Ermina Begovic [1,*], Carlo Bertorello [1], Ferdi Cakici [2], Emre Kahramanoglu [2] and Barbara Rinauro [1]

[1] Department of Industrial Engineering, University of Naples Federico II, 80125 Napoli, Italy; bertorel@unina.it (C.B.); barbara.rinauro@unina.it (B.R.)

[2] Department of Naval Architecture and Marine Engineering, Yildiz Technical University, 34349 Istanbul, Turkey; fcakici@yildiz.edu.tr (F.C.); emrek@yildiz.edu.tr (E.K.)

* Correspondence: begovic@unina.it; Tel.: +39-081-768-3708

Received: 23 March 2020; Accepted: 4 May 2020; Published: 9 May 2020

Abstract: In this paper, the validation of the hybrid frequency–time domain method for the assessment of hard chine displacement hull from vertical motions is presented. Excitation and hydrodynamic coefficients in regular waves are obtained from the 3D panel method by Hydrostar® software, while coupled heave and pitch motions are calculated in the time domain by applying the Cummins equations. Experiments using a 1:15 scale model of a "low-drag" small craft are performed in irregular head and following waves at Froude numbers Fr: 0.2, 0.4, and 0.6 at University of Naples Federico II, Italy. Results obtained by hybrid frequency–time domain simulations for heave, pitch, and vertical accelerations at center of gravity and bow are compared with experimental data and showed high accuracy.

Keywords: Cummins equations; vertical motions assessment; time domain simulations; experimental seakeeping; hard chine displacement hull form

---

1. Introduction

The assessment of ship behavior in an irregular seaway is one of the most difficult hydrodynamic problems. A large variety of different computational methods have been presented in the past three decades and are discussed in Hirdais et al. [1]. They proposed the subdivision into six levels, where each "level" introduces mathematical models closer to the physical models, generally moving from frequency domain calculations to time domain simulations, from linearized to nonlinear boundary conditions, and from small wave amplitudes to breaking waves, spray, and water flowing onto and off the ship's deck.

The simplest "Level 1" approach considers the potential flow linearized frequency domain methods, while "Level 6" deals with fully non-linear methods like Reynolds Averaged Navier Stokes (RANS) and Smooth Particle Hydrodynamics (SPH). Significant simplifications have been introduced in mathematical models considering separately vertical and horizontal motions and neglecting the viscous effects and the ship's transversal symmetry.

Two main approaches are known—the frequency and time domains. Frequency domain methods have been widely used as strip theories (2D) or as panel methods (3D) with different levels of nonlinearities considered in the mathematical model. They are valid under small wave amplitudes and small ship motions hypothesis, assuming linearized boundary conditions and the linear superposition principle. If the phenomenon involves non-linearities, rising from high wave amplitudes, high advancing speed, or hull forms with strong flare, then the time domain approach should be considered. Today, time domain simulations based on Cummins equation [2] are becoming standard as they can accommodate nonlinearities due to nonlinear wetted surface and steeper waves and are very fast. The original Cummins' equation considers the fluid memory effects for radiation terms that are calculated using the damping and added mass values in the frequency domain. Although the direct calculation of convolution integrals in Cummins equation is possible (so called "full" time domain calculation), as stated in Perez and Fossen [3], this is very time consuming, and for analysis and control system design, the convolutions are not suitable. One of the possible solutions is the introduction of "parametric model identification" since the convolution is a dynamic linear operator and can be represented by a linear ordinary differential equation-state-space model or transfer functions in the Laplace domain [3]. Perez and Fossen [4] schematized works on parametric model identification developed during time as Time Domain Identification (TDI) and Frequency Domain Identification (FDI) and reported the major contributions by different authors. Armesto et al. [5] presented results of time domain simulations for the motion of the water inside an oscillating water column and a free decay test in the heave of a spar buoy by three techniques—the direct solution of convolution integral, an approximation of the convolution integral with state space model, and Prony's estimation of the convolution. The state space method and Prony's approximations are computationally cheaper, and their results are very close to those from the direct solution of convolution integrals. The authors reported different uncertainties seen in the identification of the coefficients in these methods and concluded that, with the increase of computational capabilities given by actual computers, they recommend the use of a direct integration method to compute the radiation term in Cummins' equation.

Some of the important works for the "Level 2" methodology validation and application cases presented in the last years are:

- Rodrigues and Guedes Soares [6] with adaptive mesh and pressure integration scheme developed concerning the hydrostatic and Froude–Krylov forces evaluation on instantaneous wet hull surface;
- Acanfora et al. [7], where "level 2" seakeeping code method has been used for the excessive accelerations calculations acting on container stacks;
- Cakici et al. [8], who presented the seakeeping results of stabilized motor yacht obtained by the hybrid method using strip theory and direct calculation of convolution integrated into the control loop;
- Kucukdemiral et al. [9] used a time domain identification method as they develop a model predictive controller for vertical motions of a passenger ship;
- Gaebele et al. [10] presented a state space model for restricted heave motion of a full-scale array of floating oscillating water column wave energy converters.

The present work is further contribution to the validations of the time domain simulations. The 2 DOF time domain simulation code, developed by the authors of this paper and explained in detail in Cakici et al. [8], has been validated for vertical motions in irregular head and following waves of the hard chine displacement boat and compared against experimental data obtained by the authors. The calculation of excitation and radiation terms in the frequency domain has been performed by Hydrostar® software based on the 3D panel methods. Direct computation of the convolution integrals has been used for fluid memory effects. Restoring terms are considered linear and are calculated from the ship main properties. The wave loads in the time domain are obtained by a realization technique from the Hydrostar® frequency domain calculations.

The considered ship is representative of actual small craft trends, oriented to the low drag simple hard chine hull form studied in Bertorello and Begovic [11]. Seakeeping characteristics of hard chine and warped hull at high speed are characterized by strong nonlinearities due to the dynamic trim and constant changing of wetted surface. The general approach for planing hulls vertical motions prediction follows time domain simulation by Zarnick's theory [12] based on the potential flow, full planing condition, and constant deadrise angle in regular waves. On the other hand, flow separation and spray formation due to the hard chine present a complex hydrodynamic problem, which can be correctly studied only by RANS methods and sophisticated mesh modelling techniques with high computational efforts. At the moment, for a warped hard chine hull operating in displacement and semi-displacement regimes, there is no adequate numerical tool. Thorough validation of the applied method has been performed considering a very demanding hull form tested in irregular head and following seas at Froude numbers Fr = 0.2, 0.4, and 0.6. Simulations are performed with the time step equal to the frequency of sampling, and an identical analysis of the numerical and experimental time series is performed to obtain the fair comparison.

2. Mathematical Model

A brief overview of the fundamental equations used in the mathematical model for vertical motions of ship advancing in irregular waves is given.

2.1. Cummins Equation for Coupled Heave and Pitch Motion

Starting from the general Cummins' Equation (1),

$$\left(\overline{M} + A^{\infty}\right)\ddot{\eta}(t) + \int_0^t K(t - \tau)\eta(\tau)d\tau + C\eta(t) = F_w(t) \tag{1}$$

the following parameters can be defined:

A^{∞} is the added mass at infinite frequency,

K is the impulse response (retardation) function matrix, defined as

$$K(t) = \frac{2}{\pi} \int_0^{\infty} [B(\omega)] \cos(\omega t)d\omega \tag{2}$$

$B(\omega)$ is the damping matrix in frequency domain.

C is the restoring matrix.

$\eta(t)$ is the oscillatory response of the ship.

$F_w(t)$ is the transient wave force vector that can be created by a linear superposition of frequency domain results for different wave spectra.

In this study, the Cummins' equation is solved for the coupled vertical motions of the ship advancing with the constant speed V, as written in Equations (3) and (4). Subscripts 3 and 5 refer to heave and pitch motions, respectively.

$$\left(\Delta + A_{33}^{\infty}\right)\ddot{\eta}_3(t) + B_{33}(\infty)\eta_3(t) + \int_0^t K_{33}(t - \tau)\eta_3(\tau)d\tau + [C_{33} + C_{33C}]\eta_3(t) + A_{35}^{\infty}\ddot{\eta}_5(t) +$$
$$B_{35}(\infty)\eta_5(t) + \int_0^t K_{35}(t - \tau)\eta_5(\tau)d\tau + [C_{35} + C_{35C}]\eta_5(t) = F_3(t) \tag{3}$$

$$\left(I_5 + A_{55}^{\infty}\right)\ddot{\eta}_5(t) + B_{55}(\infty)\eta_5(t) + \int_0^t K_{55}(t - \tau)\eta_5(\tau)d\tau + [C_{55} + C_{55C}]\eta_5(t) + A_{53}^{\infty}\ddot{\eta}_3(t)$$
$$+ B_{53}(\infty)\eta_3(t) + \int_0^t K_{53}(t - \tau)\eta_3(\tau)d\tau + [C_{53} + C_{53C}]\eta_3(t) = F_5(t) \tag{4}$$

Added mass values at infinite frequency are dependent on the ship geometry only and can be obtained as the convergence value from the frequency domain added mass graphs. Bij(V) term stands for the constant damping arises from forward speed of the ship. According to Riemann–Lesbesque, Lemma Bij(V) can be replaced with B(∞) [4,13,14]. Restoring coefficients C_{33}, C_{35}, C_{53}, and C_{55} represent the constant values and they are calculated from ship main properties.

Convolution term for the restoring coefficients C_{33C}, C_{35C}, C_{53C}, and C_{55C} are also called radiation restoring terms, represent the correction to the hydrodynamic steady forces acting upon the ship surface and can be calculated as follows:

$$C_{ijC} = \omega_e{}^2\left[A_{ij}^\infty - A_{ij}(\omega_e)\right] - \omega_e \int_0^\infty \left[K_{ij}(t)\sin\omega_e t\right]dt, \ i, \ j = 3, \ 5 \tag{5}$$

If regular waves are considered, as in Fonseca and Soares [15], then the C_{ijC} will be constant and independent of the encounter frequency. While restoring terms are depending on the ship geometry and mass distribution and are independent of wave frequencies, the radiation restoring coefficient is introduced to accommodate for a correction to the hydrostatic buoyancy force/moments due to the unsteady ship motions. In Fonseca and Soares [16–18] and in Vásquez et al. [19], the radiation restoring and the memory functions are obtained by relating the radiation forces in the time domain and in the frequency domain by means of Fourier analysis. Ma et al. [20] theoretically derived a formulation for radiation restoring coefficient by use of strip theory, and for the considered test cases, it seemed the more consistent formulation. Since the irregular waves are considered in the present study, the effects of radiation restoring are neglected.

2.2. Calculation of Convolution Terms

$K_{ij}(t)$ is the impulse response function, which can be calculated by Equation (6) for the coupled heave and pitch motions as:

$$K_{ij}(t) = \frac{2}{\pi} \int_0^\infty \left[B_{ij}(\omega_e) - B_{ij}(\infty)\right]\cos(\omega_e t)d\omega_e \tag{6}$$

where $i, j = 3, 5$.

The damping values $B_{ij}(\omega_e)$ are found using the 3D panel method implemented in Hydrostar® software by Bureau Veritas, France. $B_{ij}(\infty)$ can be obtained as the convergence value from the graphs of damping coefficients as function of the wave frequency. In the present study, three forward speeds are considered, and for each speed, the encounter frequencies are calculated in the range of wave frequencies $\omega = 0.3$ rad/s to 2.1 rad/s. The definite integrals for impulse response functions defined by Equation (6) are calculated for head waves since the encounter frequencies range are sufficiently large, and it covers the following wave frequencies as well. It is noted that once the impulse response functions are calculated for three forward speeds for head waves, these values can also be used for the following wave simulations.

For calculation of the convolution integrals in Equations (3) and (4), the following approximation is applied. If the convolution integrals are split into two additional parts, one can obtain the following statements:

$$X_1 = \int_0^t K_{33}(t-\tau)\eta_3(\tau)d\tau + \int_0^t K_{35}(t-\tau)\eta_5(\tau)d\tau \tag{7a}$$

$$X_2 = \int_0^t K_{53}(t-\tau)\,\eta_3(\tau)d\tau + \int_0^t K_{55}(t-\tau)\,\eta_5(\tau)d\tau \tag{7b}$$

As noted in Cakici et al. [8], Equations (7a) and (7b) can be approximated as [21–23]

$$X_1 = \sum_{n=0}^{N} K_{33}(n\,\Delta t)\,\eta_3(t-n\Delta t)\Delta t + \sum_{n=0}^{N} K_{35}(n\,\Delta t)\,\eta_5(t-n\Delta t)\Delta t \tag{7c}$$

$$X_2 = \sum_{n=0}^{N} K_{53}(n\,\Delta t)\eta_3(t - n\,\Delta t)\Delta t + \sum_{n=0}^{N} K_{55}(n\,\Delta t)\eta_5(t - n\,\Delta t)\Delta t \tag{7d}$$

where $N = \frac{t^1}{\Delta t}$, Δt is the time step size t^1 maximum time value,

The rule of thumb for choosing the time step and the maximum time value is given by the report of McTaggart [24], formulating

$$\Delta t \approx 0.05 \sqrt{\frac{L}{g}} \text{ and } t^1 \approx 5 \sqrt{\frac{L}{g}}$$

These formulations suggest that the time interval Δt should be sufficiently small to capture the variation of retardation function, and the maximum time value t^1 should encompass the time when the retardation functions approach zero. In any case, the maximum time value should be determined according to retardation function plots.

2.3. Heave Force and Pitch Moment Calculation

To obtain randomized time record of heave force $F_3(x,t)$ and pitch moment $F_5(x,t)$, first the response spectra of heave force and pitch moment $S_{HF/PM}(\omega_E)$ are calculated, using the linear superposition principle proposed by St Dennis and Pierson [25] as in Equation (8):

$$S_{HF/PM}(\omega_E) = S_\zeta(\omega_E) \times |TF_{HF/PM}(\omega_E, \chi)|^2 \tag{8}$$

where

$S_\zeta(\omega_E)$ is the encounter wave energy spectrum, in this work JONSWAP spectrum was used;
$TF_{HF/PM}(\omega_E, \chi)$ is the transfer function (RAO) of heave force or pitch moment;
χ denotes the encounter angle;
$S_{HF/PM}(\omega_E)$ is the heave force or pitch moment response spectrum.

According to linear random wave theory, unidirectional and long-crested waves can be expressed by the sum of finite regular wave components. The instantaneous heave force amplitude can be stated as follows:

$$F_3(x,t) = \sum_{i}^{N} F_{3i} \cos(\omega_{ei}t + \varepsilon_i + \arg[F_{3i}(\omega_{ei}, \chi)]) \tag{9}$$

$F_{3i} = \sqrt{2S_{HF}(\omega_{ei})\Delta\omega_{ei}}$ represents the amplitude of i-th heave force component,
ω_{ei} is the i-th encounter wave frequency;
ε_i is the i-th phase lag, randomly assigned between 0 and 2π. It is noted that once random phase lag is chosen, it will be the same for pitch moment;
$\arg[F_{3i}(\omega_{ei}, \chi)]$ denotes heave force phase angle.

2.4. State Space Representation of Mathematical Model of Vertical Ship Motions

To solve the system of equations, the state space representation is used as explained in the following Equation:

$$\frac{d}{dt}\eta(t) = A\eta(t) + B_1[\text{wave}(t) - \text{convo}(t)] \tag{10}$$

where $\eta(t) \in \mathfrak{R}^n$ is the differentiable state vector, $\text{wave}(t) \in \mathfrak{R}^{m_W}$ is the wave load input vector, and $\text{convo}(t) \in \mathfrak{R}^{m_C}$ is the convolution vector. A, B_1 are known appropriate state space matrices.

The system matrices, disturbance inputs and states of the system are defined as

$$
A = \begin{bmatrix} 0 & 0 & 1 & 1 \\ 0 & 0 & 1 & 1 \\ -M^{-1}C & -M^{-1}B \end{bmatrix}, \; B_1 = \begin{bmatrix} 0 & 0 \\ 0 & 0 \\ M^{-1}\begin{bmatrix} 1 & 0 \\ 0 & 1 \end{bmatrix} \end{bmatrix},
$$

$$
\text{wave}(t) = \begin{bmatrix} F_3 \\ F_5 \end{bmatrix}, \; \text{convo}(t) = \begin{bmatrix} X_1 \\ X_2 \end{bmatrix}, \; \eta(t) = \begin{bmatrix} \eta_3 \\ \eta_5 \\ \eta_3 \\ \eta_5 \end{bmatrix}
$$

The statements in the state space matrices are defined as follows:

$$
M = \begin{bmatrix} \Delta + A_{33}^{\infty} & A_{35}^{\infty} \\ A_{53}^{\infty} & I_{55} + A_{55}^{\infty} \end{bmatrix}, \; C = \begin{bmatrix} C_{33} & C_{35} \\ C_{53} & C_{55} \end{bmatrix}, \; B = \begin{bmatrix} B_{33}^{\infty} & B_{35}^{\infty} \\ B_{53}^{\infty} & B_{55}^{\infty} \end{bmatrix}
$$

3. Experimental Campaign for Hard Chine Displacement Hull Form

3.1. Experimental Setup and Model Description

The experimental campaign was performed in the Towing Tank of University of Naples Federico II (UNINA), Italy. The towing tank dimensions are $135 \times 9 \times 4.2$ m and it has a wave generator capable of generating waves from 0.20 /hz to 1.25 Hz and towing carriage with maximum speed of 8 m/s. The simulations are performed for the hard chine displacement hull form, studied in Bertorello and Begovic [11] and shown in Figure 1.

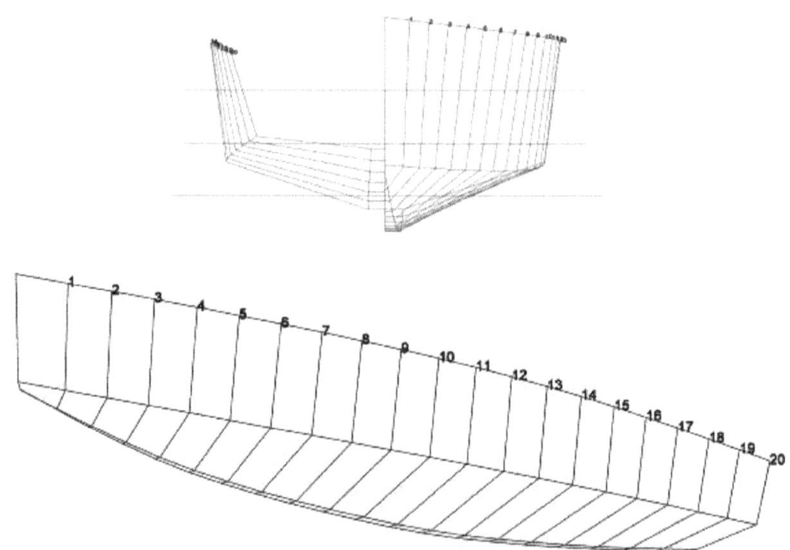

Figure 1. Hull form from Bertorello and Begovic [11].

A wooden model of 2.00 m L_{OA} was tested at one displacement (342.17 N) in irregular waves, completing the previously performed experimental campaign in calm water and regular waves in Bertorello and Begovic [11]. The values of main characteristics for the tested model are given in Table 1.

Table 1. The main characteristics of the tested model.

Parameter	Model	Ship
Scale factor $\lambda = 15$		
L_{OA} (m)	2.000	30.000
L_{WL} (m)	1.994	29.910
B_{WL} (m)	0.494	7.410
T (m)	0.116	1.740
Δ (N,kN)	342.17	1183.7
LCG (m)	0.932	13.980
VCG (m)	0.204	3.060
r_{55} (kgm^2)	0.487	7.300
Static trim (deg)	0.000	0.000
Pos. of the accelerometer at bow (m) (from CoG)	0.963	14.445

The model has been ballasted again, and the center of gravity position and moments of inertia were measured by the model inertial balance shown in Figure 2.

Figure 2. The model on the inertial balance at the Towing Tank of University of Naples Federico II (UNINA).

The towing point was at x = 0.87 m from the stern, and the weight of the mechanical arm is the part of the model ballast. The model scale ratio is $\lambda = 15$, which corresponds to 30 m L_{OA}.

Measurement of pitch and heave motions were performed by the mechanical arm R47, which holds the model restrained to surge, sway, roll, and yaw. Accelerations were measured at the CG and at the bow. Three ultrasonic wave gauges were used for the wave measurements. Two wave gauges were aligned in the front of the model at the distance of 1.93 m from the R47, and one was aligned with the R47 on the tank side. All data are sampled at a frequency of 500 Hz without filtering. The experimental set up is shown in Figure 3.

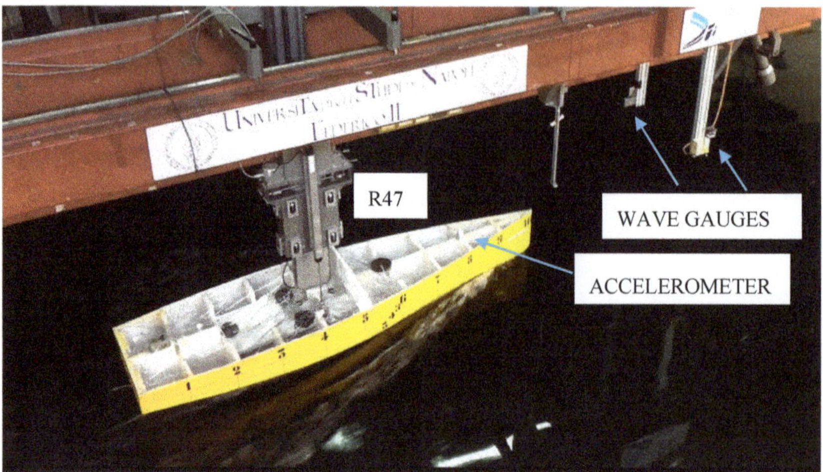

Figure 3. Experimental setup of low drag hard chine hull at UNINA.

3.2. Experiments in Irregular Waves

The experiments in irregular head and following waves were performed to validate the hybrid frequency–time domain numerical method. The standard JONSWAP theoretical spectra, defined as

$$S_{\varsigma\text{-}JONSWAP}(\omega) = A_\gamma S_{PM}(\omega)\gamma^{\exp\left(\frac{-1}{2}\left(\frac{\omega-\omega_P}{\sigma\omega_P}\right)^2\right)} \tag{11}$$

where

S_{PM}—Pierson-Moskowitz spectrum, as defined by DNV-RP-C205 [26];
γ—non-dimensional peak shape parameter;
σ—spectral width parameter, $\sigma = 0.07$ for $\omega \leq \omega_P$, $\sigma = 0.09$ for $\omega > \omega_P$;
$A\gamma$—normalizing factor, defined as $A_\gamma = 1 - 0.287\ln(\gamma)$;

were used for the representation of irregular waves. To obtain enough encounters, the runs were repeated 2–10 times, depending on the model speed and heading. For the head sea, the complete time series is 140 s, resulting in a minimum of 120 wave encounters. In the following sea, due to the very low (or negative) encounter frequency, the total time of experimental series is around 200 s, ending up with a minimum of 30 encounters. The JONSWAP spectrum parameters for model and ship scale are reported in Table 2. The example of the measured spectrum is given in Figure 4, together with the ideal wave and encounter spectra. Experimental set up and tested model velocities were identical to those in regular waves. In Table 3, the number of encounters, significant wave height, and m0 are given for the spectra measured at the following three Froude numbers: 0.2, 0.4, and 0.6 in head waves.

Table 2. Wave spectrum characteristics.

	SI Unit	Model Scale	Ship Scale
Significant wave height Hs	(m)	0.096	1.44
Peak period Tp	(s)	1.429	5.533
Peakness parameter γ	(-)	3.3	3.3

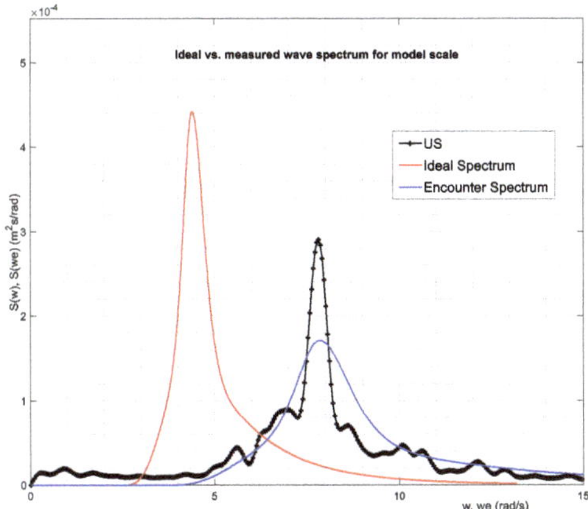

Figure 4. Ideal wave and encounter vs. measured JONSWAP spectra at Froude number Fr = 0.4.

Table 3. Measured spectra properties-head waves.

	Fr = 0.2	Fr = 0.4	Fr = 0.6
Number of encounters	144	167	219
Significant wave height Hs (m)	0.098	0.096	0.097
Moment of 0 order (m²)	0.00060	0.00058	0.00059
Total Runs Time (s)	140	140	140

Examples of the time series of 140 s registration of the wave, heave, and pitch at Fr = 0.4 in the head and following seas are given in Figures 5a and 6. These show the appreciated different number of wave encounters, and consequently of heave and pitch oscillations for the same registration time in the head and following sea. In Figure 5b, measured accelerations at center of gravity CG and at the bow are given. In the following waves, the measured accelerations were very low, and the signals were noisy, so for the sake of clarity, these data have been neglected for the further comparisons. It can be noted from Figure 5b that the accelerations are normalized by g, so they oscillate around 1g value.

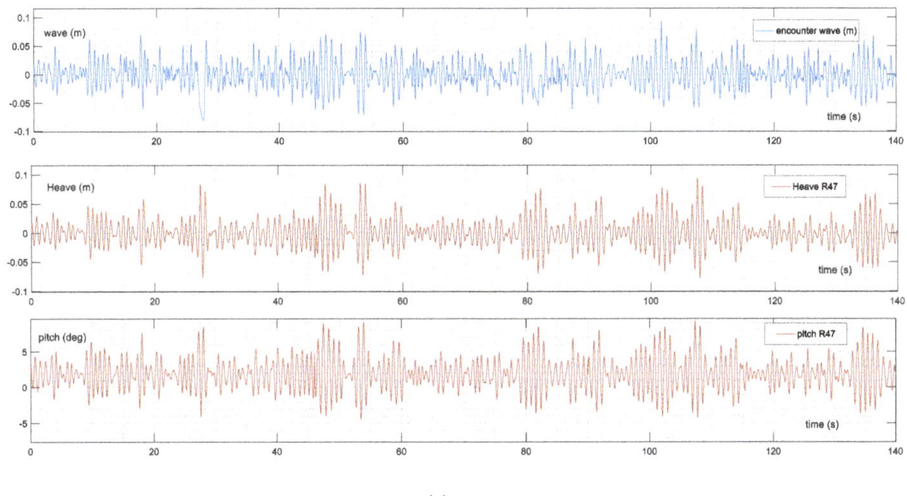

(**a**)

Figure 5. *Cont.*

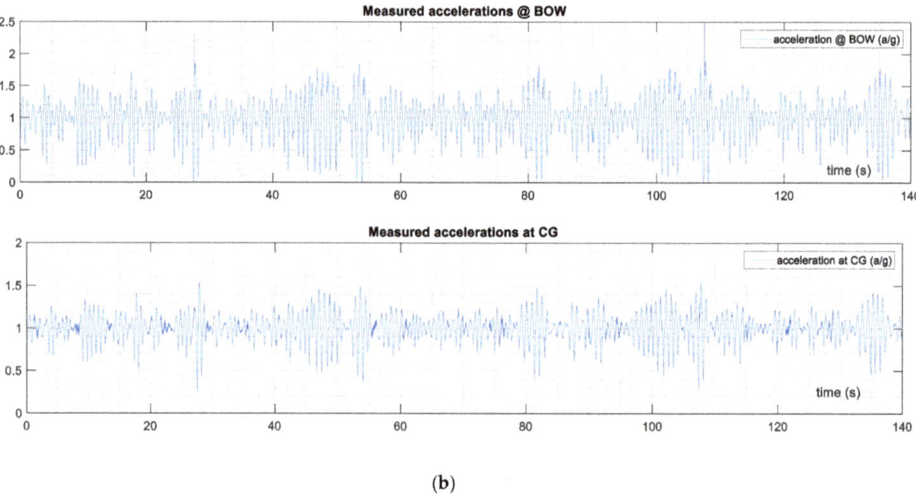

(b)

Figure 5. (**a**) Example of measured wave, heave, and pitch at Fr = 0.4 in head waves; (**b**) example of measured accelerations at Fr = 0.4 in head waves.

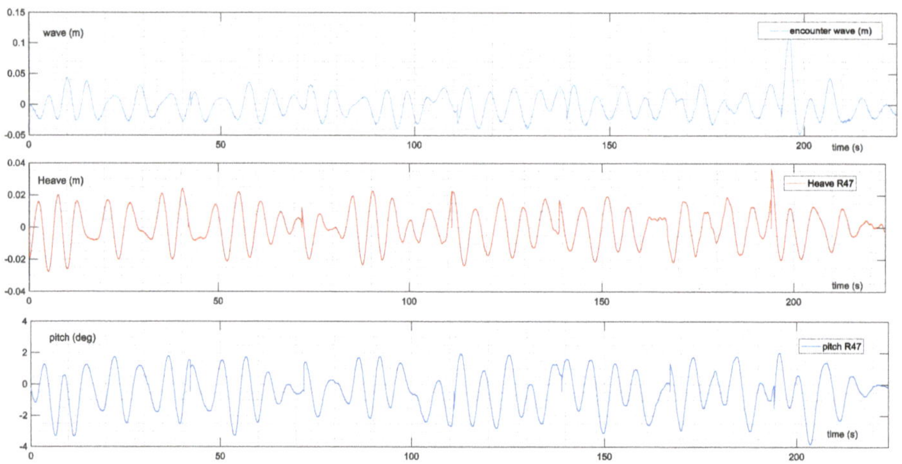

Figure 6. Example of measured wave, heave, and pitch at Fr = 0.4 in the following waves.

4. Hydrodynamic Coefficients and Exciting Forces Calculation

To be able to calculate motions in the time domain, the hydrodynamic coefficients for added mass and damping and exciting forces are needed as input, and they were obtained by the hydrodynamic software HydroSTAR® developed by Bureau Veritas.

HydroStar® is a 3D diffraction/radiation potential theory software based on the Green function method for wave–body interactions, and it provides a complete solution of first- and second-order wave loads. Calculations have been performed in ship scale for the following three Froude number cases: 0.2, 0.4, and 0.6, which correspond to ship speed values of 3.385, 6.77, and 10.155 m/s. For each of the Froude number cases, the frequencies of forward incoming regular waves have been considered in the range from 0.3 to rad/s 2.1 rad/s, with the step of 0.05, corresponding to wave periods of 2 s to 20 s.

The panelized hull geometry representation of the model in HydroSTAR® is shown in Figure 7.

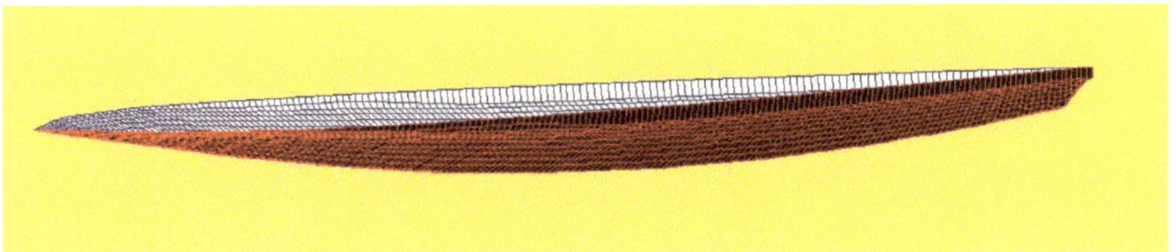

Figure 7. Hull geometry in Hydrostar® software.

For the simulation in the time domain, the following hydrodynamic coefficients and transfer functions TF values have been calculated:

- Added mass coefficients A_{33}, A_{35}, A_{55}, A_{53} in Figure 8;
- Damping coefficients B_{33}, B_{35}, B_{55}, B_{53} for the ship reported as functions of the wave frequencies ω (rad/s) in Figure 9;
- Heave and Pitch exciting forces for the ship reported in Figure 10.

All diagrams show ship values as a function of wave frequencies ω (rad/s). The model values, used as input for the time domain simulations, are obtained by applying the scaling law as indicated in Table 4 and taking values at the highest wave frequencies.

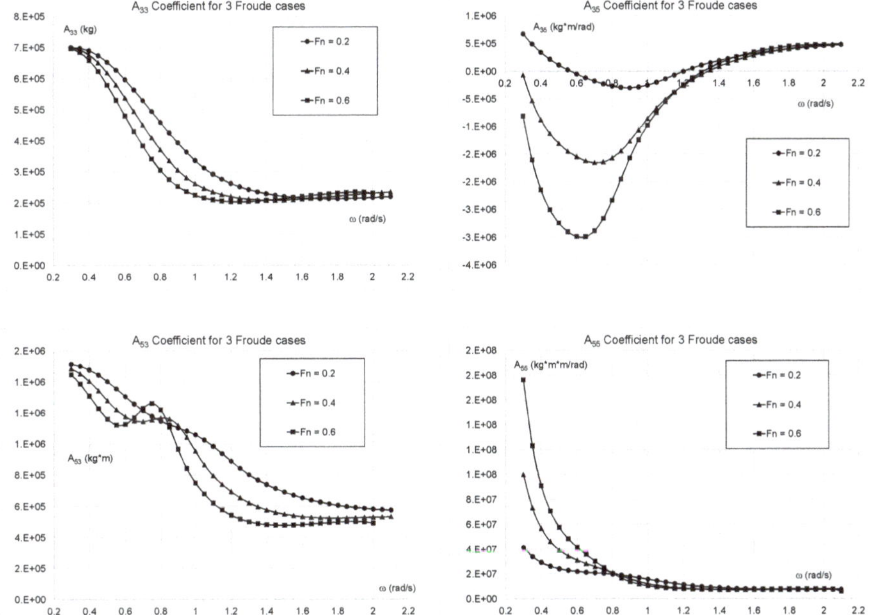

Figure 8. Added mass coefficients in the range of considered wave frequencies.

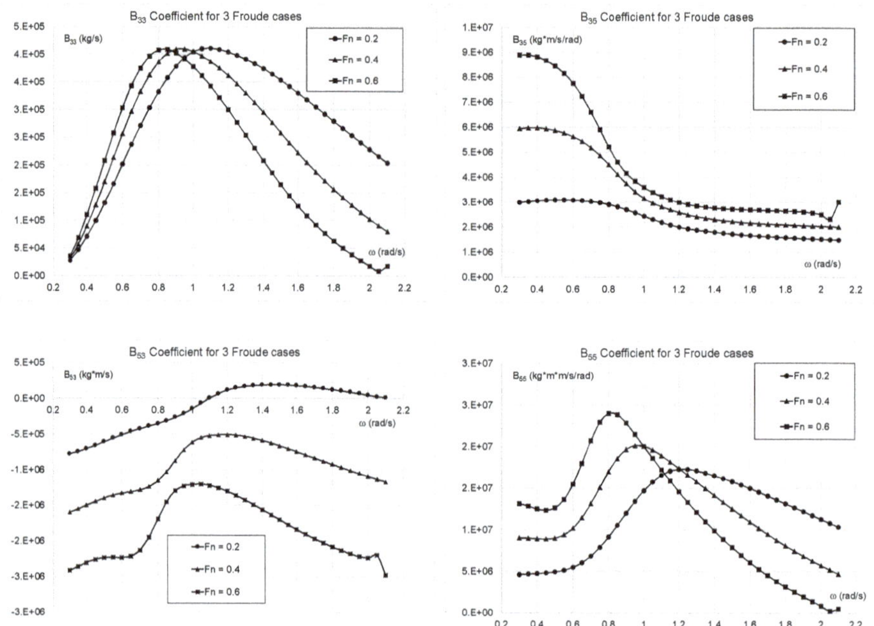

Figure 9. Damping coefficients in the range of considered wave frequencies.

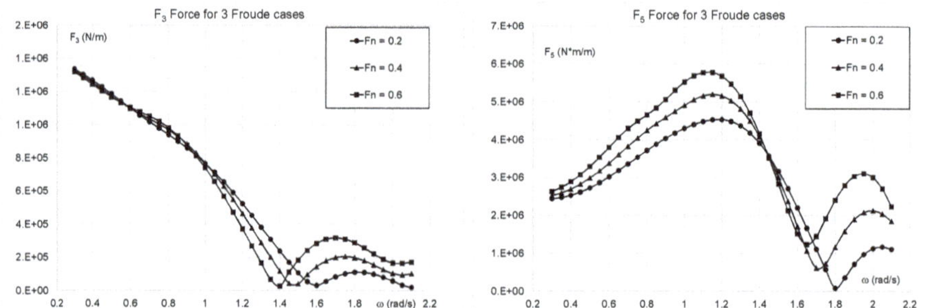

Figure 10. Transfer function (TF) of heave and pitch exciting force.

5. Time Domain Simulations

5.1. Numerical Results

The simulations with the developed code are performed in model scale to be able to compare the time series directly with the measured data. The added mass and restoring coefficients from Hydrostar® are recalculated in model scale as reported in the Table 4. The values of the added mass coefficients at infinity used in further calculations for retardation functions are reported in the last column of the Table 4. While at the sufficiently high frequencies, added mass coefficients are converging to the same value for all Froude numbers, the values of the damping coefficients at infinity are dependent on the forward speed of the model. In Table 4, for the sake of compactness, only the values for Fr = 0.4 are reported.

Retardation function has been calculated for 5 s, with the time step equal to 0.002 s, and an example at Fr = 0.4 is reported in Figure 11. Details of the 30 s of time series of heave force and pitch moment acting on the hull surface calculated according to the realization technique, in head and following waves, are given in Figures 12 and 13.

Table 4. Scaling of hydrodynamic coefficients from Hydrostar® to model scale.

Parameter	SI	Scale Factor	Model Scale
A_{33}^{∞}	(kg)	λ^3	65.185
A_{35}^{∞}	(kg*m/rad)	λ^4	9.442
A_{53}^{∞}	(kg*m)	λ^4	11.338
A_{55}^{∞}	(kg*m²/rad)	λ^5	9.653
B_{33}^{∞}	(kg/s)	$\lambda^{2.5}$	91.230
B_{35}^{∞}	(kg*m/(rad*s))	$\lambda^{3.5}$	153.00
B_{53}^{∞}	(kg*m/s)	$\lambda^{3.5}$	−89.510
B_{55}^{∞}	(kg*m²/(rad*s²))	$\lambda^{4.5}$	23.820
C_{33}	(kg/s²)	λ^2	6.356×10^3
C_{35}	(kg*m/(rad*s²))	λ^3	7.526×10^2
C_{53}	(kg*m/s²)	λ^3	7.526×10^2
C_{55}	(kg*m²/(rad*s²))	λ^4	1.258×10^3

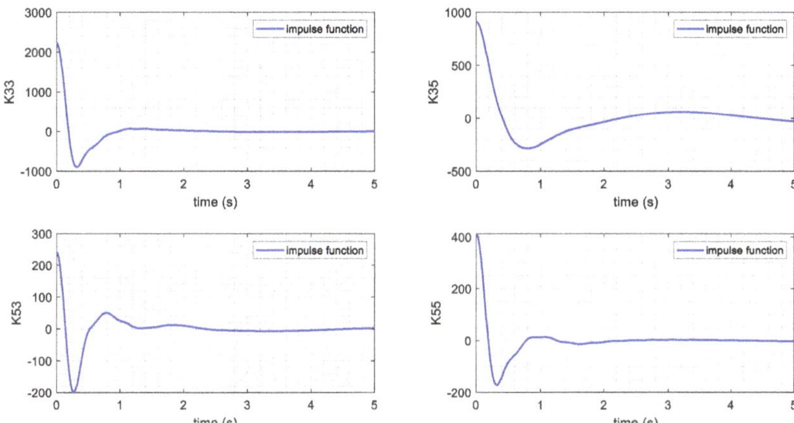

Figure 11. Impulse response functions for the Fr = 0.4.

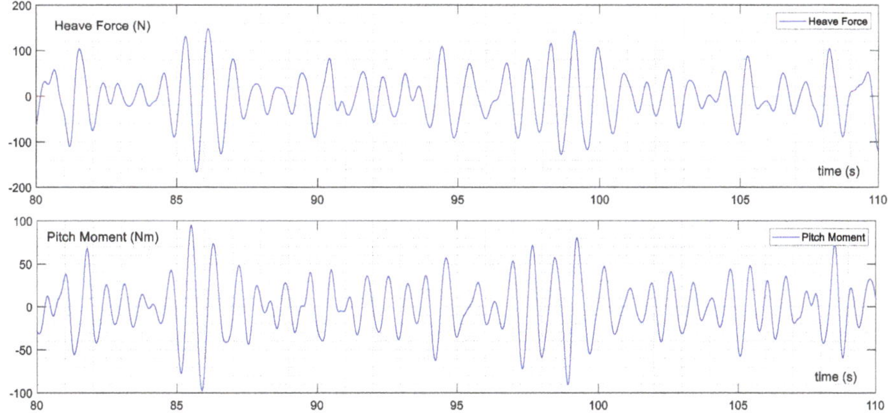

Figure 12. Detail of wave excitations time series for the Fr = 0.4 in head sea.

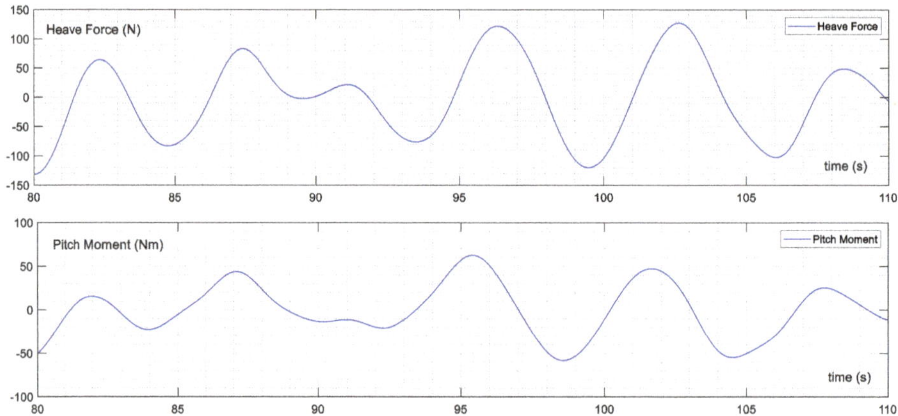

Figure 13. Detail of wave excitation time series for the Fr = 0.4 in following sea.

For the solution, The fourth-order Runge Kutta Method is used, where the time step size is taken as 0.002, and the simulation time has been set equal to 140 s for head waves and 280 s for following waves to have a fair comparison with the experimental data. In Figures 14–19, the examples of comparison of 30 s of simulated and measured time histories of motions and accelerations are given. It has to be noted that this comparison has to be seen only as a qualitative control of simulated data. Both time series are random processes and are given for the same interval of time, without trying to overlap signals. Therefore, they can be compared only statistically.

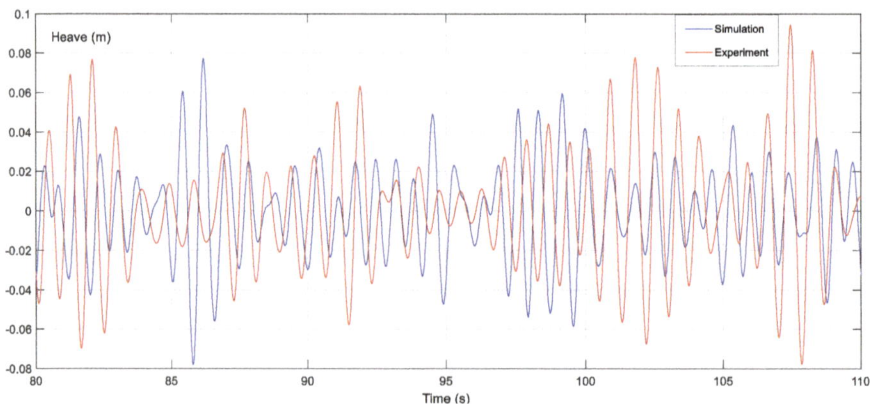

Figure 14. Example of simulated and measured heave motion at Fr = 0.4 in head seas.

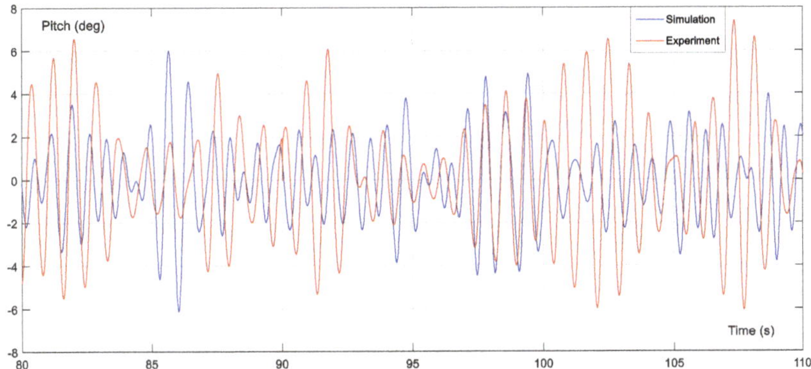

Figure 15. Example of simulated and measured pitch motion at Fr = 0.4.

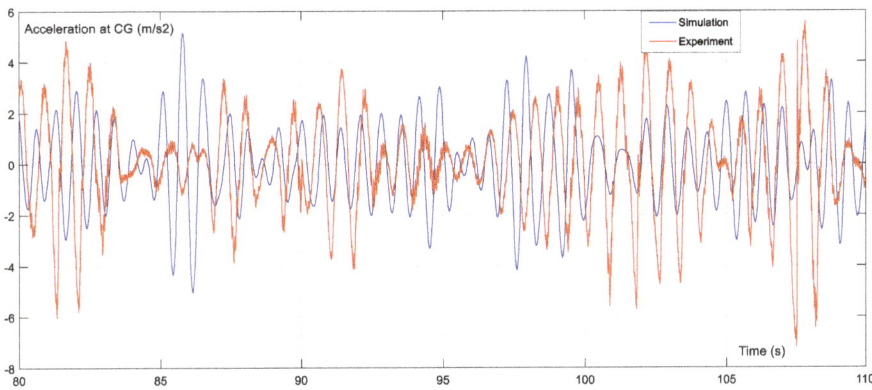

Figure 16. Example of simulated and measured cg accelerations at Fr = 0.4 in head seas.

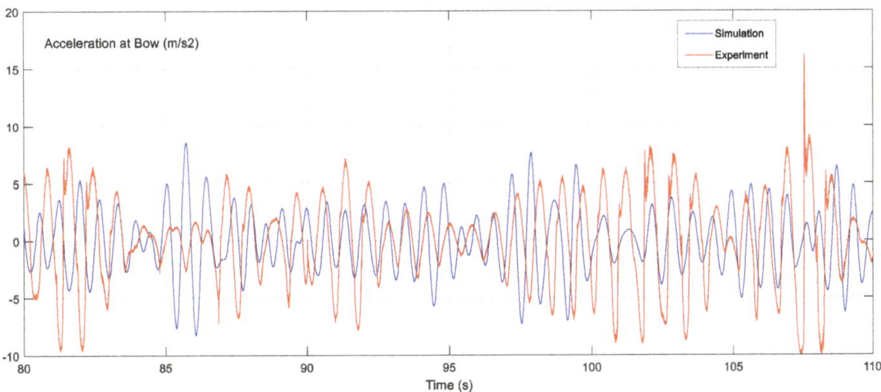

Figure 17. Example of simulated and measured bow accelerations at Fr = 0.4 in head seas.

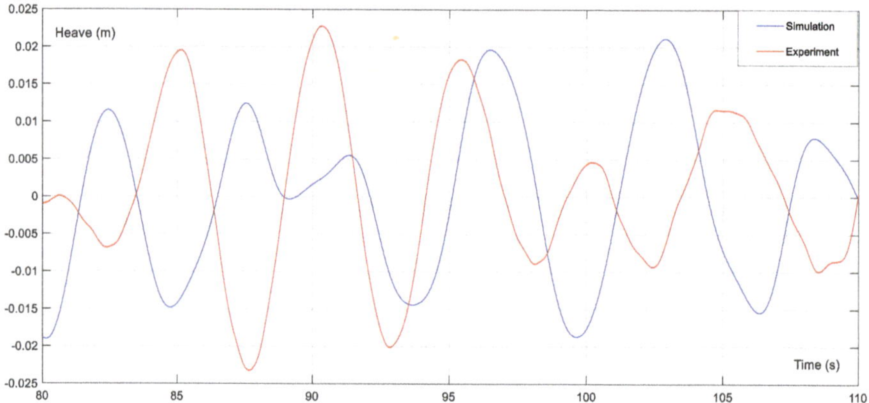

Figure 18. Example of simulated and measured heave motion at Fr = 0.4 in following seas.

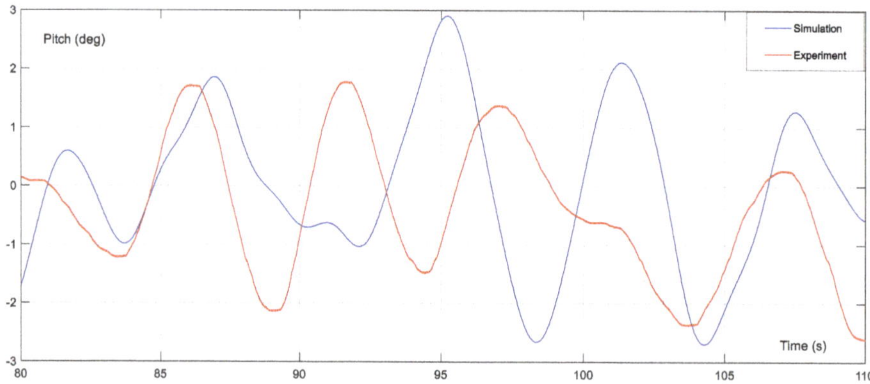

Figure 19. Example of simulated and measured pitch motion at Fr = 0.4 in following seas.

5.2. Results Comparison

Both numerical and experimental results have been analyzed in the time domain "peak to peak analysis." The complete set of performed calculations and experiments are summarized in Table 5. Tables 6 and 7 report the number of peaks in each time series, significant height, 1/10th of the highest value, maximum height value, and mean values. The differences among mean values have been expressed in the last column. It can be noted from Table 7 that calculated vertical accelerations in the following waves are very low, in the same order of magnitude as the measured ones. As the analysis of measured accelerations is strongly related to the filtering technique and cut off frequencies for the case of following waves, they have not been considered in the comparison.

Table 5. Summary of performed experiments and simulations.

No.	V	β	Fr	T_P	f_P	ω_P	ω_{EP}	Sim Time	Exp Time
	(m/s)	(deg)	(-)	(s)	(Hz)	(rad/s)	(rad/s)	(s)	(s)
1	0.879	180	0.2	1.429	0.700	4.397	6.130	140	140
2	1.758	180	0.4	1.429	0.700	4.397	7.863	140	140
3	2.635	180	0.6	1.429	0.700	4.397	9.592	140	140
4	0.879	0	0.2	1.429	0.700	4.397	2.664	280	230
5	1.758	0	0.4	1.429	0.700	4.397	0.931	280	224
6	2.635	0	0.6	1.429	0.700	4.397	−0.798	280	191

Table 6. Head seas results comparison.

| | | NUMERICAL_Fr = 0.2 | | | | | EXPERIMENTAL_Fr = 0.2 | | | | | DIFF (%) |
		No.	$H_{1/3}$	H_{MEAN}	$H_{1/10}$	H_{MAX}	No.	$H_{1/3}$	H_{MEAN}	$H_{1/10}$	H_{MAX}	H_{MEAN}
Heave	(m)	144	0.084	0.053	0.010	0.010	118	0.065	0.060	0.119	0.170	−11.4
Pitch	(deg)	145	9.417	6.085	11.19	11.189	125	9.962	6.763	13.737	17.676	−10.0
acc_cg	(g)	137	0.383	0.262	0.441	0.441	145	0.324	0.204	0.447	0.651	28.2
acc_bow	(g)	143	0.852	0.576	0.997	0.997	148	0.875	0.556	1.158	1.521	3.4
		NUMERICAL_Fr = 0.4					EXPERIMENTAL_Fr = 0.4					DIFF (%)
		No.	$H_{1/3}$	H_{MEAN}	$H_{1/10}$	H_{MAX}	No.	$H_{1/3}$	H_{MEAN}	$H_{1/10}$	H_{MAX}	H_{MEAN}
Heave	(m)	170	0.102	0.071	0.123	0.123	142	0.102	0.065	0.133	0.171	9.1
Pitch	(deg)	186	8.131	5.376	10.18	10.178	149	9.535	6.236	12.008	13.747	−13.8
acc_cg	(g)	167	0.751	0.513	0.929	0.929	167	0.750	0.476	1.001	1.161	7.8
acc_bow	(g)	173	1.305	0.8760	1.627	1.627	160	1.382	0.922	1.744	2.563	−5.0
		NUMERICAL_Fr = 0.6					EXPERIMENTAL_Fr = 0.6					DIFF (%)
		No.	$H_{1/3}$	H_{MEAN}	$H_{1/10}$	H_{MAX}	No.	$H_{1/3}$	H_{MEAN}	$H_{1/10}$	H_{MAX}	H_{MEAN}
Heave	(m)	218	0.094	0.059	0.121	0.121	173	0.083	0.055	0.110	0.132	7.3
Pitch	(deg)	240	6.405	4.337	8.223	8.223	179	5.890	3.950	7.779	9.680	9.8
acc_cg	(g)	219	0.897	0.612	1.173	1.173	197	0.846	0.562	1.123	1.444	8.7
acc_bow	(g)	230	1.407	0.987	1.794	1.794	191	1.261	0.834	1.727	2.568	18.3

Table 7. Following seas results comparison.

| | | NUMERICAL_Fr = 0.2 | | | | | EXPERIMENTAL_Fr = 0.2 | | | | | DIFF (%) |
		No.	$H_{1/3}$	H_{MEAN}	$H_{1/10}$	H_{MAX}	No.	$H_{1/3}$	H_{MEAN}	$H_{1/10}$	H_{MAX}	H_{MEAN}
heave	(m)	114	0.044	0.028	0.056	0.066	81	0.051	0.033	0.069	0.081	14.6
Pitch	(deg)	117	5.688	3.602	6.730	7.722	89	5.386	3.566	7.365	8.752	−1.0
acc_cg	(g)	107	0.031	0.021	0.037	0.044	70					
acc_bow	(g)	102	0.059	0.041	0.069	0.077	84					
		NUMERICAL_Fr = 0.4					EXPERIMENTAL_Fr = 04					DIFF (%)
		No.	$H_{1/3}$	H_{MEAN}	$H_{1/10}$	H_{MAX}	No.	$H_{1/3}$	H_{MEAN}	$H_{1/10}$	H_{MAX}	H_{MEAN}
heave	(m)	50	0.037	0.024	0.043	0.049	37	0.042	0.032	0.048	0.057	24.0
pitch	(deg)	45	4.303	2.953	5.008	5.562	34	4.444	3.516	4.740	4.801	16.5
acc_cg	(g)	75	0.005	0.004	0.007	0.009						
acc_bow	(g)	78	0.008	0.005	0.010	0.012						
		NUMERICAL_Fr = 0.6					EXPERIMENTAL_Fr = 0.6					DIFF (%)
		No.	$H_{1/3}$	H_{MEAN}	$H_{1/10}$	H_{MAX}	No.	$H_{1/3}$	H_{MEAN}	$H_{1/10}$	H_{MAX}	H_{MEAN}
heave	(m)	120	0.049	0.034	0.059	0.068	38	0.041	0.030	0.044	0.044	−13.4
pitch	(deg)	133	5.324	3.549	6.534	7.392	33	4.840	3.316	5.488	6.096	−7.0
acc_cg	(g)	213	0.050	0.033	0.060	0.074						
acc_bow	(g)	244	0.115	0.078	0.142	0.168						

In Tables 6 and 7 the comparison for head and following seas, respectively, are given.

As deduced from Table 6, the differences between numerical and experimental heave-pitch motions at head waves in terms of significant mean values remain in the range of 10% for all tested forward speeds, which indicates that the predicted motions have great level of accuracy when it is compared to experimental data. Only for the accelerations at CG at Fr = 0.2 and bow accelerations at Fr = 0.6 differences are 28% and 18%, respectively.

As shown in Table 7, in the case of following waves, the differences between numerical and experimental heave-pitch motions are around 10%, except for Fr = 0.4, where the difference is around 24%. It has to be noted that at this Fr, encounter frequencies are close to zero, which is well known situation where there are theoretical restrictions on damping coefficients. Since this Fr also corresponds to the case in which wave celerity and ship speed are the same, it is expected that linear tools might fail to represent this case. As stated previously, in following waves, both measured and calculated vertical accelerations are very low, in the order of magnitude 0.02g, and therefore, this comparison has not been considered.

6. Conclusions

In the present work, a hybrid frequency–time domain 2 DOF simulations code is presented. The prediction of the wave loads in the time domain was made by a realization technique from the linear frequency domain calculations by HydroStar® software. Direct computation of the convolution integrals was used for the fluid memory effects. The validation of the method was performed comparing the numerical results with the experimental data for a hard chine displacement hull at three model speeds in irregular head and following waves.

Results showed that the applied method for the prediction of vertical ship motions in the time domain is reliable for head and following wave cases. The accuracy of the proposed numerical implementation is dependent on the calculation of the impulse response function, which is, in turn, a function of damping values in the frequency domain. As underlined, a 3D panel method is used in this work. If strip theory is used for the evaluation of damping terms, attention must be taken because there are theoretical restrictions at the low-frequency region that directly affect the impulse response function signal. In this situation, filtering for the frequency dependent damping function may help despite the loss of accuracy. Very challenging validation of the developed method has been performed considering hard chine warped displacement hull, relative ship speed corresponding up to Fr = 0.6, and the following sea at the low and negative encounter frequencies. It has to be highlighted that the calculations in HydroStar® have been performed at the static trim, although at the Fr = 0.6 the dynamic trim is around 1.5 deg and this has been neglected in the calculation of the hydrodynamic coefficients. Obtained results in terms of percentage difference of the mean values is at max 15% in all cases, except in following sea at a model speed close to the wave celerity. In that case, the differences are around 25%. For the following sea, the shorter times of the experimental series and the obvious benefit in using the numerical tool, reasonably accurate, and fast to obtain vertical motions in the time domain in the design procedure should all be noted. Future work will consider the implementation and validation of six DOF methodologies.

Author Contributions: Conceptualization, data curation, investigation, methodology, validation, writing—original draft, writing—review and editing, supervision: E.B.; Conceptualization, Data Curation, Investigation, Writing—Original Draft: C.B.; Conceptualization, data curation, methodology, software, validation, writing—original draft, writing—review and editing: F.C.; Formal analysis, software, validation, writing—review and editing: E.K.; Data curation, investigation, writing—review and editing: B.R. All authors have read and agreed to the published version of the manuscript.

Funding: Ferdi Çakici and Emre Kahramanoğlu were supported by the Scientific and Technological Research Council of Turkey (TÜBİTAK).

Acknowledgments: The calculation by HydroSTAR® software were possible thanks to the cooperation agreement between Bureau Veritas and the Department of Industrial Engineering of University of Naples Federico II (https://www.docenti.unina.it/webdocenti-be/allegati/materiale-didattico/576434).

Conflicts of Interest: The authors declare no conflict of interest.

References

1. Hirdaris, S.E.; Bai, W.; Dessi, D.; Ergin, A.; Gu, X.; Hermundstad, O.A.; Huijsmans, R.; Iijima, K.; Nielsen, U.D.; Parunov, J.; et al. Loads for use in the design of ships and offshore structures. *Ocean Eng.* **2014**, *78*, 131–174. [CrossRef]
2. Cummins, W.E. The impulse response function and ship motions. *Schiffstechnik* **1962**, *9*, 101–109.
3. Perez, T.; Fossen, T. A matlab toolbox for parametric identification of radiation force models of ships and offshore structures. *Model. Identif. Control* **2009**, *30*, 1–15. [CrossRef]
4. Perez, T.; Fossen, T. Parametric Time Domain models based on frequency domain data. In *One-Day Tutorial, CAMS'07*; A Norwegian Research Bulletin: Bol, Croatia, 2007.
5. Armesto, J.A.; Guanche, R.; Jesus, F.; Iturrioz, A.; Losada, I.J. Comparative analysis of the methods to compute the radiation terms in Cummins' Equation. *J. Ocean Eng. Mar. Energy* **2015**, 377–393. [CrossRef]

6.   Rodrigues, J.M.; Guedes Soares, C. Froude-Krylov forces from exact pressure integrations on adaptive panel meshes in a time domain partially nonlinear model for ship motions. *Ocean Eng.* **2017**, *139*, 169–183. [CrossRef]
7.   Acanfora, M.; Montewka, J.; Hintz, T.; Matusiak, J. On the estimations of of the design loads on container stacks due to excessive acceleration in adverse weather conditions. *Mar. Struct.* **2017**, *53*, 105–123. [CrossRef]
8.   Cakici, F.; Yazici, H.; Alkan, A.D. Optimal control design for reducing vertical acceleration of a motor yacht form. *Ocean Eng.* **2018**, *169*, 636–650. [CrossRef]
9.   Kucukdemiral, I.B.; Cakici, F.; Yazici, H. A model predictive vertical motion control of a passenger ship. *Ocean Eng.* **2019**, 186. [CrossRef]
10.   Gaebele, D.T.; Magana, M.E.; Brekken, T.K.A.; Sawodny, O. State space model of an array of oscillating water column wave energy converters with inter-body hydrodynamic coupling. *Ocean Eng.* **2020**, 195. [CrossRef]
11.   Begovic, E.; Bertorello, C. Low drag hull form for medium large yachts. In *Design and Construction of Super & Mega Yachts*; RINA: Genoa, Italy, 2017; pp. 81–89.
12.   Zarnick, E. *A Nonlinear Mathematical Model of Motions of a Planing Boat in Regular Waves*; David W Taylor Naval Ship Research and Development Center: Bethesda, MD, USA, 1978.
13.   Fossen, T.I. Environmental Forces and Moments. In *Lecture Notes TTK 4190 Guidance and Control of Vehicles*; NTNU: Trondheim, Norway, 2013; Chapter 8.
14.   Perez, T. *Ship Motion Control. Course Keeping and Roll Stabilisation Using Rudder and Fins*; Springer: New York, NY, USA, 2005.
15.   Fonseca, N.; Soares, G.C. *Time Domain Analysis of Vertical Ship Motions, Transactions on the Built Environment*; Wit Press: Southampton, UK, 1994; Volume 5, ISSN 1743-3509.
16.   Fonseca, N.; Guedes Soares, C. Comparison between experimental and numerical results of the nonlinear vertical ship motions and loads on a containership in regular waves. *Int. Shipbuild. Prog.* **2005**, *52*, 57–89.
17.   Fonseca, N.; Guedes Soares, C. Comparison of numerical and experimental results of nonlinear wave-induced vertical ship motions and loads. *J. Mar. Sci. Technol.* **2002**, *6*, 193–204. [CrossRef]
18.   Fonseca, N.; Guedes Soares, C. Time-domain analysis and wave loads of large-amplitude vertical ship motions. *J. Ship Res.* **1998**, *42*, 139–153.
19.   Vásquez, G.; Fonseca, N.; Guedes Soares, C. Experimental and numerical study of the vertical motions of a bulk carrier and a Ro–Ro ship in extreme waves. *J. Ocean Eng. Mar. Energy* **2015**, *1*, 237–253. [CrossRef]
20.   Ma, S.; Wang, R.; Zhang, J.; Duan, W.; Cengiz Ertekind, R.; Chen, X. Consistent formulation of ship motions in time-domain simulations by use of the results of the strip theory. *Ship Technol. Res.* **2016**. [CrossRef]
21.   Kurniawan, A.; Hals, J.; Moan, T. Assessments of time-domain models of wave energy conversion system. In Proceedings of the Ninth European Wave and Tidal Energy Conference, Southampton, UK, 5–9 September 2011.
22.   Kashiwagi, M. Transient responses of a VLFS during landing and take-off of an airplane. *J. Mar. Sci. Technol.* **2004**, *9*, 14–23. [CrossRef]
23.   Taghipour, R.; Perez, T.; Moan, T. Hybrid frequency-time domain models for dynamic response analysis of marine structures. *Ocean Eng.* **2008**, *35*, 685–705. [CrossRef]
24.   McTaggart, K. ShipMo3D Version 3.0 user manual for creating ship models. In *Technical Memorandum DRDC Atlantic TM 2011-307*; Defence Research and Development Canada: Ottawa, ON, Canada, 2011.
25.   St Denis, M.; Pierson, W. On the motions of ships in confused seas. *Trans. Soc. Nav. Arch. Mar. Eng.* **1953**, *61*, 280–354.
26.   Det Norske Veritas. *Recommended Practice DNV-RP-C205*; Det Norske Veritas: Oslo, Norway, 2014.

Journal of
Marine Science and Engineering

Article

IMO Second Generation Intact Stability Criteria: General Overview and Focus on Operational Measures

Nicola Petacco * and Paola Gualeni

Department of Electric, Electronic and Telecommunication Engineering and Naval Architecture (DITEN), University of Genova, 16126 Genova, Italy; paola.gualeni@unige.it
* Correspondence: nicola.petacco@edu.unige.it

Received: 10 June 2020; Accepted: 3 July 2020; Published: 5 July 2020

Abstract: At the beginning of 2020, after a long and demanding process, the Second Generation Intact Stability criteria (SGISc) have been finalized at the 7th session of the International Maritime Organization (IMO) sub-committee on Ship Design and Construction (SDC). At present, SGISc are not mandatory, nevertheless IMO endorses their application in order to assess their consistency and validity. It is envisaged that SGISc can support the design of safer ships, nevertheless such a rules framework might have an impact also on the ship operational aspects in a seaway. In fact, within the SGISc framework, Operational Measures have also been implemented providing guidance and limitations during navigation. After a comprehensive overview about SGISc vulnerability levels and direct stability assessment, this paper provides a specific insight into the methodological approach for the Operational Measures extensively addressed as a complementary action to ship design.

Keywords: second generation intact stability criteria; operational guidance; operational limitations; vulnerability levels; direct stability assessment; Ro-Ro ferry

---

1. Introduction

The so called Second Generation Intact Stability criteria (SGISc) have been finalized during the 7th session of the IMO sub-committee on Ship Design and Construction (SDC) in 2020. Currently, the criteria are intended not to be mandatory but they received the endorsement of IMO, to be extensively applied in the shipping community. It is foreseen that they need further refinement, but it is expected that SGISc will positively influence the ship design process in the next years. The ship stability performance in waves has been addressed by the SGISc, with specific focus on five dynamic phenomena that is, parametric roll, pure loss of stability, dead ship condition, surf-riding and excessive accelerations. An interesting innovation introduced by the SGISc is the multi-layered approach which defines three assessment levels, characterized by different level of accuracy and therefore conservativeness. Adopting this structure, a designer may choose the kind of analysis to be carried out about the ship stability performance.

Moreover, within the SGISc framework, ship operational aspects during navigation have been introduced. It is recognized that, in order to get a safer ship performance, addressing only design aspect cannot be enough. Operational measures should be also taken into account and indications should be provided to the master. For this reason, both Operational Limitations and Operational Guidance have been developed in the SGISc framework.

In the following sections, an overview of the development process that lead to the finalized version of SGISc is given. Moreover, all the five stability failure modes have been presented with a brief description of the vulnerability level requirements. Finally, a focus on the Operational Measures defined by the SGISc is provided with considerations about the relevant approach and the possible acceptable scenarios.

2. From the Intact Stability Code to the Second Generation Intact Stability Criteria

For a long time, through the Rahola criteria [1] and the Weather criterion [2], until the Intact Stability (IS) code issued at the beginning of the 21st Century, the stability of intact ships in calm water has been one of the main topics in the naval architecture field and in the rule making environment. An exhaustive description of the history of intact stability criteria, from the origin up to the IS code, can be found in References [3,4].

During the discussion about the finalization of the IS code, the need to consider more attentively the physics of what may lead to stability failure has been pointed out. This approach, which is less dependant on database of previous incidents, could guarantee criteria in principle applicable regardless the ship typology, adaptable also for new ship projects but above all able to take into account also the dynamic behaviour of ships due to the presence of waves. In the discussion, the new modality has been often named as a *physical approach*, implying therefore also the possibility to take into consideration the hydrodynamic aspects of a ship in a seaway condition. Adopting such approach requires an enhanced knowledge of the complex behaviour of ship in a rough sea and of the strong non linearities that might entail even the ship capsizing phenomena [5–8].

The IS code has been finalized in 2008 and the need of performance-based criteria addressing ship stability in a seaway condition has been introduced and highlighted in its text, more precisely within the general provisions of its mandatory part [9] (Part A—Section 1.1).

The stability failure modes identified by the IS code are the ones addressed within the new intact stability criteria [10–12] and already previously taken into consideration in some IMO Circulars [13,14]:

- Restoring moment variation due to waves profile (Parametric Rolling and Pure Loss of Stability);
- Stability failure in the Dead Ship condition;
- Manoeuvring-related stability failure (Surf-Riding and Broaching-to).

In the early 2008, during the 51st session of the IMO Sub-committee on Stability and Load lines and on Fishing vessels safety (SLF), an inter-sessional correspondence group was established *ad hoc* [15] (Section 4.27) with the aim to develop a set of criteria for the above identified stability failure modes. At the 53rd session of SLF in 2010 instead of *new generation* criteria, it has been proposed the current name of criteria, that is, Second Generation Intact Stability criteria (SGISc). Moreover, the excessive acceleration stability problem was added as a further phenomenon to be addressed in the SGISc framework [16,17]. The framework and the relevant terminology have been identified along with the development of the process for the methodology and the criteria identification. Soon, the task appeared to be complex and the required time for its accomplishment long accordingly.

2.1. The Multi-Layered Approach

The so called *multi-layered approach* has been an interesting innovation introduced in the SGISc creation. It consists of a set of criteria, defined for each stability failure, characterized by an increasing level of accuracy, in relation with the layer. This approach assumes three assessment levels: the first two levels, namely the vulnerability assessment levels, are meant to identify ships which could be vulnerable to the stability failure under investigation; conversely, the third level is a purely performance-based direct stability assessment. In addition to what above, also the operational aspects of a ship have been considered within the SGISc framework and relevant assessment approaches to identify Operational Measures have been provided as well. The criteria defined at the first vulnerability level (Lv1), for each stability failure mode, are a simplified check to initially separate roughly vulnerable ships from not vulnerable ones. First vulnerability criteria consist of simple assessments, taking into account few ship geometrical data and loading condition parameters, for example, metacentric height and righting moment in calm water. Second vulnerability levels (Lv2) require the knowledge of more details than Lv1; the phenomena are addressed relying on physics-based approaches taking possibly into account the ship dynamic behaviour in a seaway condition. In the criteria of Lv2, the environmental condition are more detailed as well; therefore a wide set of sea states is processed and a sort of long-term analysis is requested to be carried out. The direct stability assessment (DSA) represents the so called third level of the multi-layered approach. It should predict as close as practically possible the actual ship motions in a seaway condition. It should consist of a non-linear time domain numerical simulations considering at least four degrees of freedom and some of their coupling factors. Model tests ensuring the same level of accuracy can be adopted as well. The DSA therefore is the most accurate, but also the most computationally time-consuming level in the SGISc framework. As a supplementary level, the SGISc introduces some measures acting on the environmental conditions and ship operational aspects—such Operational Measure (OM) respectively are named as the Operational Limitations (OL) and Operational Guidance (OG). The latter is a document containing information and recommendation about ship navigation with the aim to reduce the likelihood of failures. Differently, the OL identify restrictions to the ship operability in relation with specific geographical area and environmental conditions, with the aim to avoid stability failures.

The multi-layered structure has been introduced in the SGISc framework with the aim of avoiding unnecessary high computational burden when not necessary, that is, when ships are nor likely to be vulnerable. In the framework, the higher is the level, the more complex is the required analysis. On the other hand, as an inherent consequence, a relevant conservative safety margin is introduced at lower levels. Therefore, the sequential application of levels would be desirable—if a ship is deemed vulnerable by Lv1, the second level criteria should be used in order to understand whether it is really an issue or not; in the case where the ship is also considered vulnerable by Lv2, the DSA should be applied as the last level, as the one characterised by the best possible reliability, before the introduction of operational measures.

Nevertheless, the finalized version SGISc deem it acceptable that the user can directly apply any design assessment (Lv1, Lv2 or DSA) or operational measures option (OG or OL), without any hierarchy as the one described above. This allow the user to start the design assessment analysis from the DSA or even to directly move to the operational level and apply OL without performing any design assessment. The simplified scheme of the application logic of SGISc is given in Figure 1.

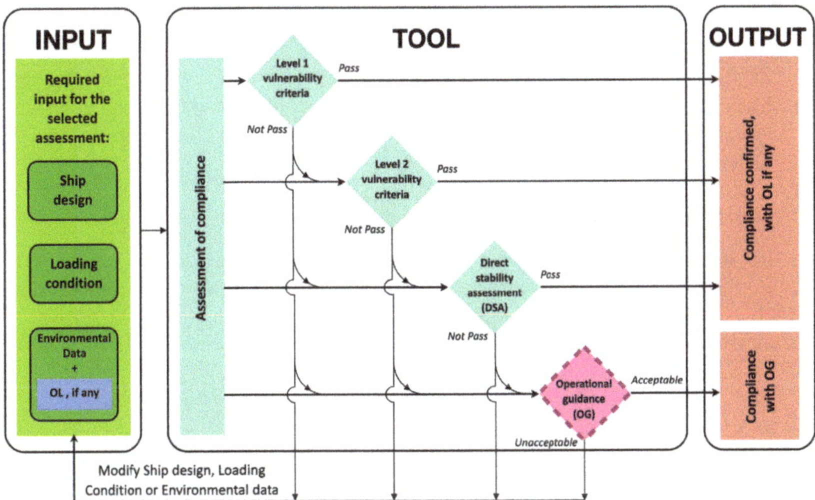

Figure 1. Representation of the simplified application scheme of the application logic in the Second Generation Intact Stability criteria (SGISc) framework.

Applying the multilayered approach in a sequential logic (from Lv1 to DSA), it is expected that the higher level gives a more reliable answer in confirming the vulnerability of the ship or in rejecting it. On the contrary, a consistency problem between levels can be identified when a lower level considers the ship not vulnerable while a higher level deems the opposite. In this perspective, during the development of the different criteria for the various vulnerability assessments, concerns about consistency of results rose, especially when Lv1 and Lv2 criteria are compared [18–21]. Most of the issues have been fixed during the refinement of the criteria formulation, even if some consistency problem still remains. In this perspective, failure modes have different evidences, for example, Lv1 and Lv2 for dead ship condition may provide non consistent results or Lv2 for pure loss of stability may appear too much conservative for ship with low freeboard. In order to bypass these consistency issues, the sequential application of the multilayered approach is not binding and the user can take advantage of the criteria support in the way it is deemed as the most appropriate.

2.2. Toward the Finalization

The SGISc development process has been articulated along the years thanks to the intense work of the inter-sessional correspondence groups and the activity during the sessions of the IMO Sub-Committees in charge (the SDC replaced the SLF after 2013). About four years since the beginning of SGISc development, the first structured draft text of the criteria has been presented at IMO [22] (Annex 1 and Annex 2). Meanwhile, a draft text also of explanatory notes has been drawn up, gathering relevant information about the physics background of each stability failure mode and the criteria formulation [23] (Annex 1–5). After this first milestone, the working group proceeded to develop the guidelines identifying the fundamental features of the numerical tool required for an appropriate DSA. The Sub Committee instructed the correspondence group to develop the DSA procedures separately for each stability failure. The first version of the Guidelines for Direct Stability Assessment have been presented at the 1st SDC session in 2014 [24] (Annex 27). Between the 5th and 6th session of SDC, the consolidated version of the vulnerability level criteria has been finalized as well as the DSA guidelines. Also the Operational Measures has been put in the final form in the same very intense period by the inter-sessional correspondence group. Eventually, as an epochal moment, the final version of the *Guidelines on the second generation intact stability criteria* has been adopted at the 7th session of SDC [25]. A further step still needed is to complete the explanatory notes; this activity is intended to be carried out within the next SDC sessions.

The SGISc have been issued as a complementary analysis aside the mandatory criteria defined in the part A of IS code. The Guidelines will be kept under review for a trial period during which the stakeholders are encouraged to provide feedback on SGISc application on existent and new ships. In the Guidelines is clearly stated that SGISc are not intended to be used instead of the existing mandatory criteria.

3. Details about Design Assessment

As stated above, SGISc address pure loss of stability (PLS), parametric rolling (PR), surf-riding (SR), dead ship condition (DS) and excessive accelerations (EA). In this section, a brief description of the physics behind each phenomenon and the requirements of each criterion is given. The final version of vulnerability levels and DSA for SGISc, as approved at 7th session of SDC, can be found in Reference [25] (Chapters 2–3).

3.1. Restoring Arm Variation Due to Waves

This phenomenon is related to the interaction between the hull geometry and the encountered wave profile. When a longitudinal wave having a length comparable to the ship length encounters a vessel, variations of waterplane area and immersed volume distribution are registered. This has an effect on the ship restoring capability. The worst case happens when the wave crest is located amidships entailing a righting arm decrease that, in comparison with the one in calm water, can be quite critical. The opposite happens when the wave trough is amidships: in this case the effect is a strengthening of the righting moment.

This phenomenon may lead to two different stability failures, that is, pure loss stability and parametric rolling. The typical scenario, that better characterizes PLS, considers a longitudinal wave as long as the vessel, approaching from stern with a celerity just faster than the ship speed. In this situation, the wave takes a long time to pass the vessel and the critical situation, with the wave crest amidship, lasts considerably decreasing the stability performance for a not negligible time.

PR can happen when the ship interacts with a train of waves having the length comparable to the ship length and an encounter period that is half of the ship natural roll period. The typical scenario for PR can be depicted as follows: when the ship is far away from its upright position and the wave trough is amidships, the righting arm is *"stronger"* and therefore the ship is vigorously pushed back to the upright position. When the ship is in the upright position, the wave crest has already moved amidship. In this condition, the righting moment is *"weaker"* and the ship rolls to an even larger heel angle on the opposite side. If the wave trough moves again amidships when the maximum heel angle on the opposite side is reached, the cycle starts again leading to heel angles larger and larger.

3.1.1. First Vulnerability Levels for PLS and PR

First levels for PLS and PR assess the metacentric height variation due to the wave profile with a very simplified formulation. Both the levels evaluate an upright hydrostatics at the draft of the specific wave trough. PR requires also an additional hydrostatics at the draft of the wave crest of the same specific wave. The wave characteristics are a length equal to the ship length and a steepness factor of $S_W = 0.0334\,(-)$ for PLS and $S_W = 0.0167\,(-)$ for PR.

The assessed loading condition of a ship is not judged vulnerable to PLS when condition (1) is verified.

$$GM_{min} \geq 0.05\,(\text{m}), \tag{1}$$

where GM_{min} is the metacentric height calculated for the hydrostatics at the drafts defined above.

The assessed loading condition of a ship is not judged vulnerable to PR when condition (2) is verified.

$$\frac{\Delta GM_1}{GM} \leq R_{PR}, \tag{2}$$

where GM is the metacentric height in calm water for the considered loading condition; ΔGM_1 is defined as $\frac{I_{T_H}-I_{T_L}}{2\cdot\nabla}$; ∇ is the immersed volume; I_{T_H} and I_{T_L} are the transverse moment of inertia of the waterplane located respectively at the drafts corresponding to the wave crest and the wave trough. The standard R_{PR} is defined as a function of breadth, length, amidship coefficient and the bilge keel projected area; it ranges from $0.17\,(-)$ to $1.87\,(-)$.

Moreover, both criteria consider a ship vulnerable if the condition (3) is not verified.

$$\frac{\nabla_D - \nabla}{A_W \cdot (D - d)} \geq 1.0. \tag{3}$$

3.1.2. Second Vulnerability Levels for PLS and PR

The structure of second vulnerability levels is very similar among all the stability failure modes addressed within the SGISc framework. A sort of long-term analysis is undertaken, based on the wave scatter diagram of North-Atlantic ocean, as shown in Equation (4). This analysis implies at first a short-term assessment for each sea state, which is different for the various stability failure modes.

$$C = \sum_{H_S}\sum_{T_Z} C_S \cdot W(H_S; T_Z). \tag{4}$$

Second vulnerability level for PLS judges a ship not vulnerable if condition (5) is verified.

$$\max(CR_1; CR_2) \leq R_{PL0}, \tag{5}$$

where $R_{PL0} = 0.06\,(-)$; CR_1 and CR_2 are the long-term indexes evaluating respectively the capsizing angle in waves and the static equilibrium heel angle under the action of a heeling lever l_{PL2} considering the restoring moment in waves. The wave to be considered are obtained by filtering the wave scatter diagram by means of the Grim's wave theory [26,27].

Second vulnerability level for PR considers a ship not vulnerable if condition (6) is verified.

$$C1 \leq R_{PR1} \quad \text{or} \quad C2 \leq R_{PR2}, \tag{6}$$

where $R_{PR1} = 0.06\,(-)$; $R_{PR2} = 0.025\,(-)$; C1 and C2 are the long-term indexes. The first long-term index takes into account the actual GM variation in waves evaluated for 16 different waves defined within the criterion. The long-term index C2 needs a 1-DoF model able to reproduce roll motions when the ship interacts with longitudinal waves. The waves required by the latter analysis are obtained as specified in the second level criteria for PLS.

3.2. Manoeuvring-Related Stability Failure

Manoeuvring-related stability problems are those referring to the broaching-to and surf-riding stability failure modes. These phenomena are strictly linked, in fact, broaching is preceded by surf-riding as studied in References [28,29].

Broaching-to can be defined as a sudden and uncontrollable turning despite the opposite action of the rudder to counter act it. This phenomenon may lead to large heel angle and even the capsizing. Instead, the surf-riding happens when a quartering waves with specific characteristics reaches the vessel and accelerates it at the wave celerity. In this condition most of the ships are directionally unstable, thus broaching-to may occur. In light of this relation between broaching and surf-riding, the developed criterion assess ship vulnerability to the latter in order to prevent broaching-to.

Surf-riding phenomenon may develop when the ship speed is comparable to the wave celerity and the wave length is about one to three times the ship length, together with a wave steepness great enough to generate a sufficient wave surge force. Surf-riding is characterized by two ship speed thresholds. A ship is affected by the surge motion when the speed is below the first threshold and SR cannot happen. In this condition, a vessel is accelerated by the front part of the approaching wave and then it is decelerated by the back part of the passing wave. Over the first speed threshold surf-riding may occur only under a specific condition. This condition depends mainly on the relative position between the ship and the wave crest. Finally, when the ship speed overreaches the second threshold, SR occurs under any condition. Due to the relative high speed of the vessel, the wave having the above mentioned characteristics trigger the surf-riding regardless the relative position between the ship and the wave crest.

3.2.1. First Vulnerability Level for SR

First vulnerability levels for SR is made up of a very simple assessment. A ship is judged not vulnerable if relations (7) are verified.

$$L \geq 200\,(\text{m}) \quad \text{or} \quad Fn \leq 0.30\,(-), \tag{7}$$

where L is the ship length and Fn is the Froude number at the service speed.

3.2.2. Second Vulnerability Level for SR

The second vulnerability level for SR adopts the same structure defined in condition (4) as a long-term assessment. A ship is judged not vulnerable to SR if condition (8) is verified.

$$C \leq R_{SR}, \tag{8}$$

with $R_{SR} = 0.005\,(-)$.

For this stability failure mode the short-term index is evaluated as defined in Equation (9).

$$C_S(H_S; T_Z) = \sum_{i=0}^{N_\lambda} \sum_{j=0}^{N_a} w_{ij} \cdot C2_{ij}, \tag{9}$$

where $w_{i,j}$ is a statistical weighting factor calculated with the joint distribution of local wave steepness and lengths; $C2_{i,j}$ is a coefficient as a function of the critical Froude number and the service ship speed. The coefficient $C2$ is evaluated by an iterative procedure where the equilibrium among ship resistance, propeller thrust and wave surge are pursued.

3.3. Dead Ship Condition

The stability problem related to the dead ship condition is already addresses by the so called *weather criterion* included in the IS code. Moreover, the MSC.1 Circular 1200 [30] has been issued by IMO during the IS code revision, introducing alternative assessment procedure for the weather criterion. All the same, it has been decided to address this stability failure mode in the SGISc framework adopting a more precise physical-based analysis.

The typical scenario of DS considers a ship which has lost its power and it is rolling under the action of wind and waves, usually turned in beam seas. The ship is assumed to be inclined leeward while rolling under the combined effect of wind and waves. In this situation, a sudden wind gust acts on the vessel when it is at the maximum windward roll angle. The criterion is aimed to assess the acceptable value of the dynamic angle of equilibrium under this circumstances.

3.3.1. First Vulnerability Level for DS

First vulnerability criterion for DS embeds the *Severe wind and rolling criterion* known as the weather criterion and defined in Reference [9] (Chapter 2). It differs only for the table of wave steepness factor, which is replaced by the extend version introduced in Reference [30]. The weather criterion is an energy-based model where the energy balance between the wind action and the righting moments is assessed.

3.3.2. Second Vulnerability Level for DS

The second vulnerability level for DS adopts a different approach with respect to the first level. Often, this is the origin of some inconsistency issues between the two levels. The Lv2 criterion relies on a dynamic-based method where the environmental condition and the ship characteristic parameters are modelled by means of a 1-DoF computational tool. This kind of analysis has to be carried out for each sea state defined in the wave scatter diagram. As a result of this dynamic assessment a short-term index is obtained.

Also in this case, a long-term analysis is carried out according to condition (4). A ship is judged not vulnerable if condition (10) is verified.

$$C \leq R_{DS0}, \tag{10}$$

where C is the long-term index and $R_{DS} = 0.06 \, (-)$ is the standard threshold.

3.4. Excessive Acceleration

The EA stability failure mode has been introduced in the SGISc assessments after a list of casualties had been submitted to IMO consideration. In particular, this problem usually involves vessels sailing in ballast condition having high values of metacentric height.

The scenario considered in the formulation of EA criteria entails the ship rolling under the action of pure beam seas in the zero-speed condition. Looking at a selected point located on board, the higher is the vertical distance from the roll axis, the longer is the shift to be covered in a half roll period.

Since the angular roll velocity is constant along the whole ship, the highest point has the fastest linear velocity in order to cover in the same time a longer distance. Since every half roll period the roll motion changes its direction, also the linear velocity direction changes and this leads to a linear transverse acceleration. The faster is the linear velocity change, the larger is the transverse acceleration due to roll motion. With reference to the close relationship between the roll period and the metacentric height, if the latter is higher the roll period is shorter and in turns this implies that large transverse acceleration may occur during roll motion.

Transverse acceleration is very dangerous onboard, both for the cargo and the crew members. Beside seasickness, it may cause loss of balance, fall or even being thrown against bulkheads or furniture. The same for the cargo, which may fall outboard or get damaged.

3.4.1. First Vulnerability Level for EA

A ship is judged not vulnerable to EA according to Lv1 if condition (11) is verified.

$$\varphi \cdot k_L \cdot \left(g + 4\pi^2 \cdot \frac{h_r}{T_{roll}^2} \right) \leq R_{EA1}, \tag{11}$$

where $R_{EA1} = 4.64 \ (\text{m/s}^2)$ is the standard; φ is the characteristic roll amplitude; k_L is a coefficient taking into account simultaneous action of roll, yaw and pitch motions; g is the gravity acceleration; T_{roll} is the roll period and h_r is the vertical distance between the roll axis and the highest point where crew or passenger may be present.

3.4.2. Second Vulnerability Level for EA

The second vulnerability level for EA requires to perform a long-term analysis as defined in condition (4). The short-term index to be considered is given in Equation (12).

$$C_S = \exp \left(-\frac{R_2^2}{2 \, \sigma_{LA}^2} \right), \tag{12}$$

where $R_2 = 9.81 \ (\text{m/s}^2)$ and σ_{LA} is the standard deviation of the lateral accelerations at zero speed in a beam sea. The short-term index C_S represents the probability to exceed a specified lateral acceleration.

3.5. Direct Stability Assessment: Requirements & Methodology

The DSA can be considered the third level of assessment and it represents the most reliable approach able to evaluate ship stability performances in a seaway condition. Nevertheless, due to the complexity of the investigated phenomenon (i.e., dynamic behaviour of a ship in a seaway), its satisfactory formulation and implementation is challenging. To this regard, significant steps forward have been made in latest years, with promising results [31,32].

The DSA has been conceived to enable the measurement of the level of safety in terms of average stability failure rate, namely the probability that the stability failure event may occur over a specified period of time. As concern the SGISc framework, the failure event is defined as either the exceedance of roll angle or the exceedance of lateral acceleration. The former is defined as the minimum value among 40°, the capsizing angle or the downfloading angle, all evaluated in calm water. The considered acceleration should be evaluated at the highest point on board the ship where crew members and/or passengers may be present, the threshold should be compliant with 9.81 (m/s^2).

The procedure for the DSA can be split in two main parts: the methodology able to reproduce reliable ship motions in a seaway, even at very large heel angles, and the post-processing procedure. Detailed specifications and requirements about DSA can be found in the *Guidelines for direct stability assessment* [25] (Chapter 3).

3.5.1. Prediction of Ship Motions

The method to adequately model and predict ship motions can be done either by a numerical tool, a model test or a combination of them. Whichever is the selected method, it should be able to reproduce the ship motions in irregular seas. Depending on the investigated stability failure, the direct assessment tool should replicate identified degrees of freedom in the time domain simulation, considering coupling factors among motions. In Table 1, required degrees of freedom to be modelled during the simulation for each stability failure mode are given.

In the guidelines, the requirements for the numerical tools are defined as well as how to model the roll damping coefficient, forces and moments acting on the hull. In order to comply with the necessary qualitative and quantitative validations, the numerical tool could be calibrated by means of model tests. The qualitative validation should ensure that the numerical tool is able to reproduce the physics of the stability failure mode considered. The quantitative validation should ensure the degree of accuracy the software is capable to reproduce the stability failure mode. As concern the prediction of ship motions, it is worth mentioning that it is sufficient to be validated only for the phenomenon considered. Therefore, the validation of numerical tool is failure-mode specific.

Table 1. Degrees of freedom required in the simulation of particular stability failure modes.

Stability Failure Mode	Surge	Sway	Heave	Roll	Pitch	Yaw	General Note
Dead Ship Condition	-	X	X	X	X	-	Aerodynamical forces should be taken into account.
Excessive Acceleration	-	-	X	X	X	-	When sway motion is not modelled, care about lateral acceleration reproduction should be paid.
Parametric Rolling	-	-	X	X	X	-	-
Pure Loss of Stability	X	X	-	X	-	X	-
Surf-Riding/Broaching-to	X	X	-	X	-	X	Hydrodynamical forces due to vortex shedding should be properly modeled.

For those degrees of freedom not included in the dynamic modelling, static equilibrium should be assumed.

3.5.2. Post-Processing of Results

Within the so called post-processing procedure, the estimation of the likelihood that a stability failure may occur is to be calculated. The *Guidelines for DSA* define three different approaches for such activity:

- Full probabilistic assessment;
- Probabilistic criteria for assumed design situation; and
- Deterministic criteria for assumed design situation.

The first method reproduces as accurately as possible the actual ship operative life, considering every combination of sea states, ship speeds and headings. The criterion is the estimate of the mean long-term stability failure rate, which is calculated as the average over all the combinations that have been simulated. All the stability failures, except for DS, should be simulated considering the heading ranging uniformly from $0°$ to $180°$ as well as the ship forward speed should be distributed from zero to the maximum speed. For the DS stability failure, simulations should be carried out considering beam seas and zero speed condition. Since stability failures may be rare, the full probabilistic method requires additional effort in the solution of the problem of rarity, especially when the mean time to failure is very long with respect to the ship natural roll period. In order to circumvent this issue, methods relying on a selection of assumed design situations have been introduced.

The assumed design situations are specifically defined for each stability failure mode, addressing the wave directions, ship speeds and wave periods to be considered in the simulation. Introducing these constrains, the amount of simulations and their duration is notably decreased. The assumed design situation can be assessed with either probabilistic or deterministic criteria. The probabilistic criteria takes into account the maximum stability failure rate over all the simulations for the considered failure mode. To reduce further the simulation time, the deterministic criterion takes into account the greatest mean three-hour maximum roll amplitude or lateral acceleration. Because of the large level of inaccuracy introduced by the deterministic procedure, an additional margin has been added in the selection of the standard: the thresholds have been halved.

In addition to the specifications and requirements for the assessment procedures described above, in Reference [25] (Chapter 3) judging criteria are reported as well, to verify whether the failure recorded during a simulation can be considered as the failure mode for which the numerical tool is validated. Since the validation of the numerical tool or model test procedure is failure-mode specific, for each phenomenon a set of different criteria is provided in the *Guidelines for DSA*.

4. An Overview on the Operational Measures

It is recognised that safety cannot only be a matter of ship design. To enhance the level of safety of a vessel in a seaway condition, an approach complementary to the ship design is required, as is also recognised in References [33,34]. This can be implemented with operational measures provided to the master, suggesting the safest behaviour in handling ship navigation for specific environmental conditions. It is worth pointing out that, however, when a vessel or a specific loading condition are affected by too many operational restrictions, it means that a refinement of the ship design is needed. Considering the framework of SGISc, the OM have been categorized in two typology — Operational Limitations and Operational Guidance.

4.1. Operational Limitations

Operational Limitations define operational limits to the navigation, for a specific loading condition, delimiting the environmental condition. The environmental condition may be limited by the OL with two restriction typologies. The first typology is related to the geographical areas where the ship sails (i.e., sheltered water, specific sea area and route) or it can be a seasonal limitation when the likelihood of stability failure is higher. This kind of limitation is called in the IMO document as *Operational limitations related to areas or routes and season*. The second OL typology is related to the significant wave height that the vessel is able to safely cope with during navigation. It is named *Operational Limitations related to maximum significant wave height* by IMO document.

OL need an assessment tool in order to be identified, for example, vulnerability levels or direct stability assessment. For this reason, OL may be considered a tool able to tune the considered operational profile in the design assessment. This concept is facilitated by the modularity structure of the SGISc which are formulated to easily introduce modifications to selected methodologies or boundary conditions. Therefore, OL should be considered as a complementary instrument to the assessment process and not as a stand-alone assessment tool. A graphical representation of this concept is proposed on Figure 2.

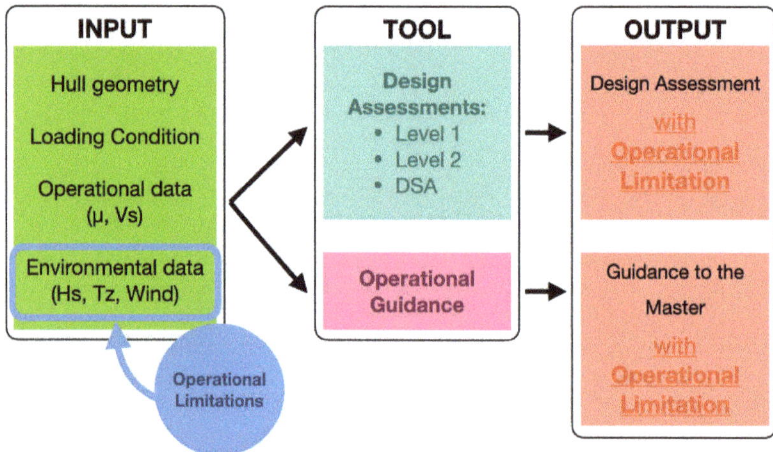

Figure 2. Scheme of the logical concept for Operational Limitations (OL).

In the SGISc, the environmental conditions can be modified by directly acting on the wave scatter diagram which is considered for the evaluation of the long term stability failure rate for each phenomena. Possible restrictions are aimed to reduce at an acceptable level the risk that a stability failure event occurs in a seaway. As regards the OL related to the geographical area, the whole scatter diagram is replaced accordingly, thus, real time weather forecasts are not required for this restrictions. On the contrary, OL related to the significant wave height modify the wave scatter diagram cutting-off sea states having a significant wave height greater than the selected threshold. The obtained scatter table is named *limited scatter table*. This limitation implies that detailed weather forecast, including significant wave heights, must be available to the master in real time, in order to avoid the limited sea states. Interesting applications of OL have been presented in References [35,36].

4.2. Operational Guidance

The Operational Guidance defines all the situations that are not recommended or should be avoided for each sea state during the navigation. An *assumed situation* represents the combination of ship operational parameters, such as ship speed and wave encounter heading (i.e., sailing condition) together with the environmental characteristics, that is, significant wave height, zero-crossing wave period, wind direction and gust characteristics. OG defines a set of operative information that support the master in the ship navigation for determined sea states. Following the suggestions of OG, the rate of a stability failure is decreased to an acceptable level. Since the OG is drawn up during the design phase, it is important that guidance addresses all the possible sea states the ship might encounter in relation with the area of navigation. As required for the OL related to the wave height, OG can be fully used only if weather forecast are available on board. In particular, detailed sea state information are beneficial for the master to plan the safest sailing condition (i.e., ship speed and heading) according to the OG.

In the SGISc framework, three different approaches have been defined in order to prepare the OG for each loading condition. They differ for the methodology which predicts the stability failure rate. According to their accuracy, an appropriately conservative threshold is introduced in order to guarantee the same level of safety. The three OG approaches are the following:

- Probabilistic operational guidance;
- Deterministic operational guidance;
- Simplified operational guidance.

The simplified OG are based on simple methodologies such as those introduced in the vulnerability levels. The other two approaches share the same method adopted in DSA, that is, a model test or a numerical calculation tool able to reproduce ship motions in the time domain with at least three degrees of freedom, considering coupling factors and necessary non linearities in irregular seas. More precise technical requirements for these methodologies are given in Reference [25] (Chapter 4).

As a consequence of the requirements briefly described above, OG can be considered an additional assessment level independent from the design assessment (i.e. Lv1, Lv2 and DSA). Although it may share the same methodology, OG can be directly performed without assessing the ship vulnerability for a specific stability failure with another assessment tool. This concept is schematically represented in Figure 3.

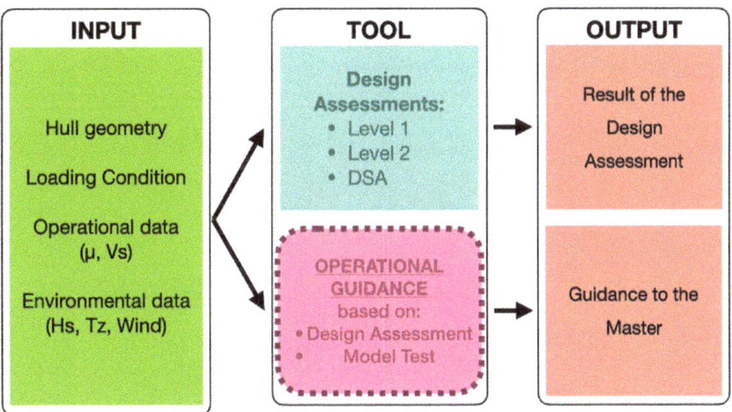

Figure 3. Scheme of the logical concept for Operational Guidance (OG).

4.3. Acceptance of Operational Measures

Because of their different concept, OG and OL can be combined together with a multiple choice of combinations. The number of possible cases increases even more when also the design assessments are taken into account. Therefore, a single loading condition can be deemed acceptable according to a set of options defined in Reference [25] (Section 4.4). These configurations have been enumerated below and gathered in Figure 4.

1. *Acceptable for unrestricted operation.* This case is applicable when the design assessments have been satisfied for each stability failure mode. Thus, no OG or OL are provided for the considered loading condition.
2. *Acceptable for limited operation.* In this case, the design assessment of one or more stability failure modes have been passed with operational limitations (e.g., limitations on wave height or on

geographical area), while the remaining stability failure have been accepted with unrestricted operation, that is, without any kind of restrictions.

3. *Acceptable for operation using onboard operational guidance.* In this configuration, the loading condition is deemed acceptable if operational guidance are provided without any restrictions for one or more stability failure, and the remaining stability failure satisfy the design assessment either with unrestricted or limited operation.

4. *Acceptable for operation in a specified area or in a specified route during a specified season.* This configuration is achieved when the design assessment for one or more stability failure mode has operational limitations related to areas or routes and season. The remaining stability failure modes do not evidence any vulnerability during the design assessment for unrestricted operation.

5. *Acceptable for limited operation in a specified area or in a specified route during a specified season.* This option allows the loading condition to have an operational limitation related to the significant wave height in a specific area or route and season for one or more stability failure modes. The remaining stability failure mode can be satisfied either with operational limitation related to the area or route and season (i.e., no limitations on the wave height) or without any other restriction nor guidance.

6. *Acceptable for operation using onboard operational guidance in a specified area or in a specified route during a specified season.* In this configuration, the loading condition is judged acceptable if operational guidance are provided together with limitations related to the area or route and season for one or more stability failure modes. The remaining stability failure can be provided either with operational limitations for the same area or route and season, otherwise without any restrictions on the design assessments.

A schematic representation of all possible cases is proposed in Figure 5. Design assessment made up of vulnerability levels (Lv1 and Lv2) and the DSA is represented. The OG and the OL are evidenced as well. The latter is divided in two sub-domains making reference to the two typologies of limitations (i.e., those related to the significant wave height and to a specific area or route and season). The number in the square stands for the possible configuration, while the arrows indicate which tools are combined in it (i.e., design assessment, operational limitation or operational guidance). Looking at the graph, it is possible to point out that only design assessment and OG may exist alone (configuration 1 and 3) while this is not true for the OL.

		Design Assessment Lv1 Lv2 DSA	Operational Guidance	Design Assessment with Operation Limitations:	
				Maximum significant wave height	Route, Season, Geographical area
Not combined		(1)	(3)	(5)	(4)
Combined with: Operational Limitations	Maximum significant wave height	(2)	(-)		(5)
	Route, Season, Geographical area		(6)	(5)	

Figure 4. Summary table of each possible configuration of Operational Measure (OM). (1) Unrestricted operation; (2) Limited operation; (3) operation using onboard OG; (4) operation in a specified area or in a specified route during a specified season; (5) limited operation in a specified area or in a specified route during a specified season; (6) operation using onboard OG in a specified area or in a specified route during a specified season.

It is important to highlight that operational guidance may indicate as safe some sailing conditions in relation to the roll motion disregarding other technical aspects, for example, limits of propulsion and steering systems, excessive vertical loads as well as slamming. This should be taken into account in order to avoid misleading OG that can jeopardise vessel navigation because of other problems. For example, sometimes transverse excessive accelerations can be reduced with increasing forward ship, but high speeds cannot be reached in some sea states or may lead to larger vertical motions and slamming.

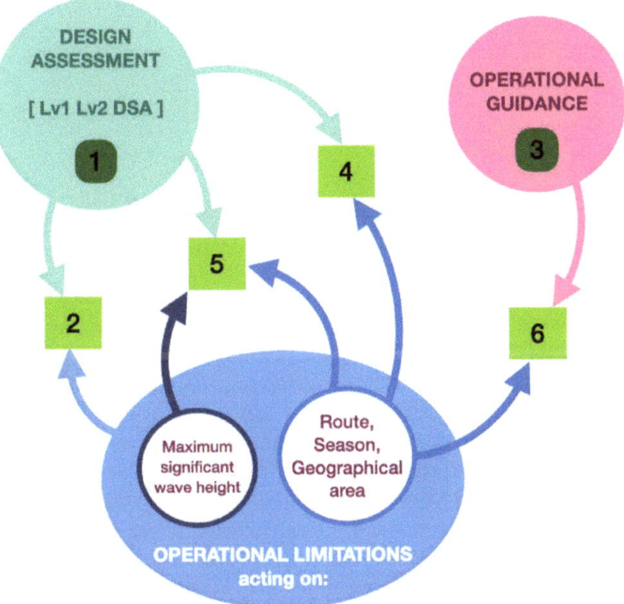

Figure 5. Graphical representation of all configuration of OM. (1) Unrestricted operation; (2) Limited operation; (3) operation using onboard OG; (4) operation in a specified area or in a specified route during a specified season; (5) limited operation in a specified area or in a specified route during a specified season; (6) operation using onboard OG in a specified area or in a specified route during a specified season.

5. An Example of Application and Results

In order to appreciate the effectiveness of vulnerability levels and operational limitations, an application to a representative Ro-Ro pax ferry is given in this section. In particular, pure loss of stability, dead ship condition and excessive acceleration have been analysed.

Principal vessel dimensions are shown in Table 2. In this analysis four drafts have been considered as representative of the most common loading conditions, that is, Departure and Arrival condition considering loaded on board only trucks or only cars. For the EA assessment, the highest point where people may be present is located in $x = 169.8$ (m) and $z = 34.0$ (m). For the DS assessment, the considered lateral exposed surface is about $A_L = 5400$ (m^2).

Table 2. Ro-Ro pax ferry main dimensions.

Main Dimensions			
Length overall	L_{OA}	211.20	(m);
Length according IS code	L_{ISc}	200.55	(m);
Maximum breadth	B_{max}	30.40	(m);
Design draught	d	7.82	(m);
Displacement	Δ	28,234	(t);
Service speed	V_S	22.0	(kt);
Block coefficient	C_B	0.650	(-);
Bilge keel length	L_{BK}	66.0	(m);
Bilge keel span	b_{BK}	0.30	(m);
Design vertical centre of gravity	KG	13.43	(m);
Design metacentric height	GM_{design}	1.43	(m).

Results are presented in terms of vertical position of center of gravity (KG) limiting curves, as shown in Figures 6–8. It is worth to note that results for EA are given by minimum KG curves. Drafts are presented on the horizontal axis, while the limiting KG are on the vertical axis. In each graph, the limiting curves obtained by the application of Lv1, Lv2, Lv2 with a limitation on the significant wave height and Lv2 with a limitation on the geographical area are shown. The first OL set an admissible significant wave height of $H_S = 7.5$ (m), while in the second limitation the unrestricted navigation has been replaced by a navigation confined in the Mediterranean sea.

The design domain is defined as the area below the maximum KG limiting curve (vice-versa for the minimum KG limiting curve due to EA), where the design centre of gravity of the ship may be placed safely. Outcomes point out that there is a significant difference in terms of design domain between first and second vulnerability levels. As expected, this is due to the caution factor inherent with the structure of first levels. Looking at the application of OL, it seems that no noticeable effect of these restrictions are evident on the second vulnerability level of DS. OL increases the design domain for PLS and EA; in particular for this vessel, the restriction on the geographical area gives the greatest improvement in term of KG domain range.

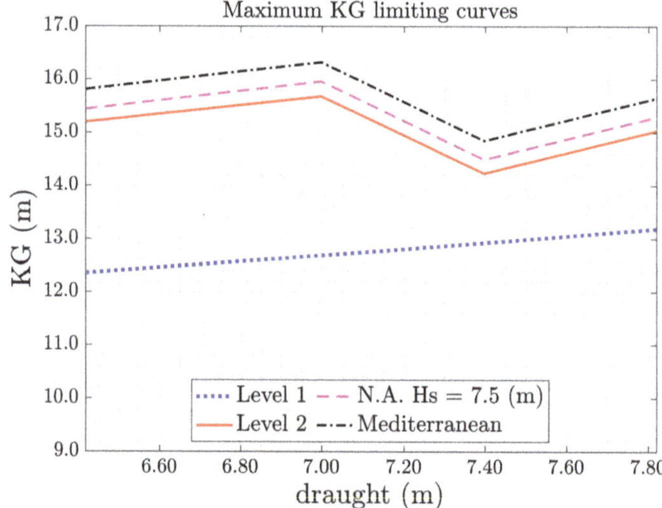

Figure 6. Vulnerability assessment in terms of maximum KG limiting curve for the pure loss of stability failure mode. Curves obtained according to the results for the application of Level 1, Level 2, Level 2 for North Atlantic Ocean limited to $H_S = 7.5$ (m) and Level 2 for Mediterranean.

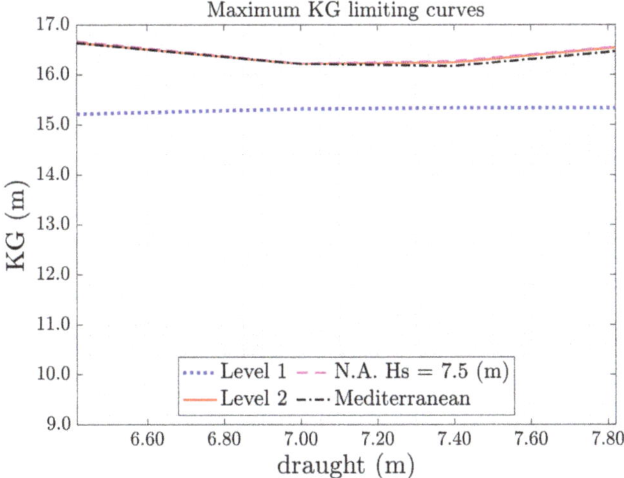

Figure 7. Vulnerability assessment in terms of maximum KG limiting curve for the dead ship failure mode. Curves obtained according to the results for the application of Level 1, Level 2, Level 2 for North Atlantic Ocean limited to $H_S = 7.5\,(\text{m})$ and Level 2 for Mediterranean.

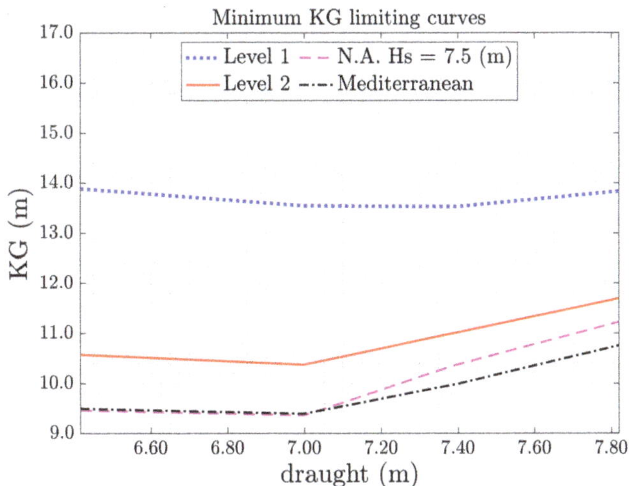

Figure 8. Vulnerability assessment in terms of minimum KG limiting curve for the excessive acceleration failure mode. Curves obtained according to the results for the application of Level 1, Level 2, Level 2 for North Atlantic Ocean limited to Hs = 7.5 (m) and Level 2 for Mediterranean.

6. Conclusions

An overview of the SGISc has been outlined, comprehensive of the first and second vulnerability level criteria (Lv1 and Lv2), the so called direct stability assessment and the Operational Measures that in turn are inclusive of Operational Limitations and Operational Guidance. Graphical representations to better describe the mutual relation among the whole set of different rules have been provided, able to deliver the complex SGISc framework at a glance. The set of rules is the results of a very innovative trade off on different domains:

- Physically based modelling and formulation;

- Computational approach and relevant computational power needed;
- Design assessment complemented with operational support and vice-versa.

The point of strength of SGISc is the ability to introduce into the stability assessment the dynamic interactions of the ship with the environment. Nevertheless, it is to be demonstrated now that such set of rules is really and effective asset to design and operate safer ships. The calculation burden and procedure complexity in some cases are really demanding and the sensitivity of results to significant design and operational parameters is to be investigated. Therefore an intensive application campaign of this innovative set of rules is encouraged by IMO in order to possibly improve the SGISc in terms of formulations, procedures and standards.

As an example, the vulnerability assessments for pure loss of stability, dead ship condition and excessive acceleration modes have been presented; operational limitations in terms of maximum significant wave height and geographical area have been further investigated as well. Outcomes show how in this case, the consistency of the multilayered philosophy has been respected between levels of the same stability failure mode. Nevertheless, the application of OL for the dead ship failure mode shows how the influence of this option is very limited compared to other stability failure modes.

For this case study, SGISc do not imply any severe constrains to the design of the vessel. The application of OL was actually not needed for the analysed vessel, although it shows that introducing restrictions—with a relatively low-impact on the operative life of a ship (i.e., maximum admissible wave height equal to 7.5 (m))—may further enlarge the design domain available to the designer. Furthermore, a proper integration of design modifications with operational measures may lead to an improvement of ship safety without any significant issue on the project. In any case, the study and definition of OG as a support to the master represents a significant aid to the navigation. It becomes an important tool to be complemented with the master experience at sea, with the aim to further improve the ship safety performance.

Author Contributions: Conceptualization, N.P. and P.G.; Software, N.P.; Supervision, P.G.; Writing—original draft, N.P.; Writing—review & editing, P.G. All authors have read and agreed to the published version of the manuscript.

Funding: This research received no external funding.

Conflicts of Interest: The authors declare no conflict of interest during preparation and publishing of this work.

Abbreviations

The following abbreviations are used in this manuscript:

DS	Dead Ship condition;
DSA	Direct Stability Assessment;
EA	Excessive Acceleration;
IMO	International Organization Maritime;
IS	Intact Stability;
Lv1	First vulnerability level;
Lv2	Second vulnerability level;
MSC	Maritime Safety Committee;
OG	Operational Guidance;
OL	Operational Limitations;
OM	Operational Measures;
PLS	Pure Loss of Stability;
PR	Parametric Rolling;
SDC	Sub-committee on Ship Design and Construction;
SGISc	Second Generation Intact Stability criteria;
SLF	Sub-committee on Stability and Load lines and on Fishing vessels safety;
SR	Surf-Riding;
1-DoF	Dynamic model with one Degree of Freedom;

References

1. Rahola, J. The Judging of the Stability of Ships and the Determination of the Minimum Amount of Stability Especially Considering the Vessels Navigating Finnish Waters. Ph.D. Thesis, Technical University of Finland, Helsinki, Finland, 1939.
2. IMO. *Recommendation on a Severe Wind and Rolling Criterion (Weather Criterion) for the Intact Stability of Passenger and Cargo Ships of 24 Metres in Length and Over*; Resolution A.562(14); International Maritime Organization: London, UK, 1982.
3. Francescutto, A. Intact stability criteria of ships—Past, present and future. *Ocean Eng.* **2016**, *120*, 312–317. [CrossRef]
4. Kobyliński, L. Stability Criteria—Present status and perspectives of improvement. *Int. J. Mar. Navig. Saf. Sea Transp.* **2014**, *8*. [CrossRef]
5. Kobyliński, L. Rational stability criteria and the probability of capsizing. In Proceedings of the 1st International Ship Stability Workshop, Glasgow, UK, 24–27 March 1975.
6. Francescutto, A. Is it really impossible to design safe ships? *Trans. R. Inst. Nav. Archit.* **1993**, *135*, 163–173.
7. Spyrou, K. Ship capsize assessment and non linear dynamics. In Proceedings of the 4th International Workshop Theoretical Advance in Ship Stability and Pratical Impact, Piraeus, Greece, 29–30 June 1998.
8. Spyrou, K.; Papanikolaou, A. Ship Design for dynamic stability. In Proceedings of the 7th International Marine Design Conference, Kyongju, Korea, 21–24 May 2000; pp. 167–178.
9. IMO. *Adoption of the International Code on Intact Stability*; Resolution MSC.267(85); International Maritime Organization: London, UK, 2008.
10. SLF 48/4/7. *Dynamic Intact Stability or Stability Problems in Waves*; Submitted by Germany; International Maritime Organization: London, UK, 2005.
11. SLF 48/WP.2. *Review of the Intact Stability Code*; Report of the Working Group (Part 1); International Maritime Organization: London, UK, 2005.
12. SLF 52/WP.1. *Development of New Generation Intact Stability Criteria*; Report of the Working Group (Part 1); International Maritime Organization: London, UK, 2010.
13. IMO. *Guidance to the Master for Avoiding Dangerous Situations in Following and Quartering Seas*; Circular MSC/707; International Maritime Organization: London, UK, 1995.
14. IMO. *Revised Guidance to the Master for Avoiding Dangerous Situations in Adverse Weather and Sea Conditions*; Circular MSC.1/1228; International Maritime Organization: London, UK, 2007.
15. SLF 52/17. *Report to the Maritime Safety Committe*; Report; International Maritime Organization: London, UK, 2008.
16. SLF 53/3/5. *Comments on the Structure of New Generation Intact Stability Criteria*; Submitted by Poland; International Maritime Organization: London, UK, 2010.
17. SLF 53/19. *Report to the Maritime Safety Committee*; Report; International Maritime Organization: London, UK, 2010.
18. SLF 53/INF.8. *Sample Calculations on the Level 2 Vulnerability Criteria for Parametric Roll*; Submitted by Sweden; International Maritime Organization: London, UK, 2010.
19. SLF 54/3/5. *Sample Verification and Proposal of Draft Level 1 Criteria on Parametric Roll and Pure Loss of Stability*; Submitted by China; International Maritime Organization: London, UK, 2011.
20. SLF 55/INF.5. *Sample Calculations for Level 1 and Level 2 Vulnerability Criteria*; Submitted by Germany; International Maritime Organization: London, UK, 2012.
21. Krüger, S.; Hatecke, H. The impact of the 2nd generation of intact stability criteria on RoRo—Ship design. In Proceedings of the 12nd International Symposium on Practical Design of Ships and Other Floating Structures, Changwon, Korea, 20–25 October 2013; pp. 641–649.
22. SLF 54/WP.3. *Development of Second Generation Intact Stability Criteria and Any Other Business*; Report of the Working Group (Part 1); International Maritime Organization: London, UK, 2012.
23. SDC 4/5/1. *Finalization of Second Generation Intact Stability Criteria—Report of the Correspondence Group*; Submitted by Japan; International Maritime Organization: London, UK, 2016.
24. SDC 1/INF.8. *Information Collected by the Correspondence Group on Intact Stability Regarding the Second Generation Intact Stability Criteria Development*; Submitted by Japan; International Maritime Organization: London, UK, 2013.

25. SDC 7/WP.6. *Finalization of Second Generation Intact Stability Criteria;* Report of the Drafting Group on Intact Stability; International Maritime Organization: London, UK, 2019.

26. Grim, O. Beitrag zu dem Problem der Sicherheit des Schiffes im Seegang. *Schiff Hafen* **1961**, *6*, 191–201.

27. Bulian, G. On an improved Grim effective wave. *Ocean Eng.* **2008**, *35*, 1811–1825. [CrossRef]

28. Umeda, N.; Matsuda, A.; Hamamoto, M.; Suzuki, S. Stability assessment for intact ships in the light of model experiments. *J. Mar. Sci. Technol.* **1999**, *45*, 45–57. [CrossRef]

29. Renilson, M. An investigation into the factors affecting the likelihood of broaching-to in following seas. In Proceedings of the 2nd International Conference on the Stability of Ships and Ocean Vehicles, Tokyo, Japan, 24–29 October 1982; pp. 551–564.

30. IMO. *Interim Guidelines for Alternative Assessment of the Weather Criterion;* Circular MSC.1/1200; International Maritime Organization: London, UK, 2006.

31. Shigunov, V.; Themelis, N.; Spyrou, K. *Contemporary Ideas on Ship Stability;* Springer: Cham, Switzerland, 2019; Volume 119, pp. 407–421.

32. Kuroda, T.; Hara, S.; Houtani, H.; Ota, D. Direct Stability Assessment for excessive acceleration failure mode and validation by model test. *Ocean Eng.* **2019**, *187*. [CrossRef]

33. Bačkalov, I.; Bulian, G.; Rosén, A.; Shigunov, V.; Themelis, N. Improvement of ship stability and safety in intact condition through operational measures: Challenges and opportunities. *Ocean Eng.* **2016**, *120*, 353–361. [CrossRef]

34. Liwång, H. Exposure, vulnerability and recoverability in relation to a ship's intact stability. *Ocean Eng.* **2019**, *187*. [CrossRef]

35. Rudaković, S.; Bačkalov, I. Operational limitations of a river-sea container vessel in the framework of the Second Generation Intact Stability Criteria. *Ocean Eng.* **2019**, *183*, 409–418. [CrossRef]

36. Petacco, N.; Gualeni, P.; Stio, G. Second Generation Intact Stability criteria: Application of operational limitations & guidance to a megayacht unit. In Proceedings of the 5th International Conference on Maritime Technology and Engineering, Lisbon, Portugal, 16–19 November 2020.

Journal of
Marine Science and Engineering

Article

Seakeeping Performance of a New Coastal Patrol Ship for the Croatian Navy

Andrija Ljulj [1] and Vedran Slapničar [2,*]

[1] Ministry of Defence of the Republic of Croatia, 10000 Zagreb, Croatia; andrija.ljulj@morh.hr
[2] Faculty of Mechanical Engineering and Naval Architecture, University of Zagreb, 10000 Zagreb, Croatia
* Correspondence: vedran.slapnicar@fsb.hr

Received: 9 June 2020; Accepted: 12 July 2020; Published: 15 July 2020

Abstract: This paper presents seakeeping test results for a coastal patrol ship (CPS) in the Croatian Navy (CN). The full-scale tests were conducted on a CPS prototype that was accepted by the CN. The seakeeping numerical prediction and model tests were done during preliminary project design. However, these results are not fully comparable with the prototype tests since the ship was lengthened in the last phases of the project. Key numerical calculations are presented. The CPS project aims to renew a part of the Croatian Coast Guard with five ships. After successful prototype acceptance trials, the Croatian Ministry of Defence (MoD) will continue building the first ship in the series in early 2020. Full-scale prototype seakeeping test results could be valuable in the design of similar CPS projects. The main aim of this paper is to publish parts of the sea trial results related to the seakeeping performance of the CPS. Coast guards around the world have numerous challenges related to peacetime tasks such as preventing human and drug trafficking, fighting terrorism, controlling immigration, and protecting the marine environmental. They must have reliable platforms with good seakeeping characteristics that are important for overall ship operations. The scientific purpose of this paper is to contribute to the design process of similar CPS projects in terms of the development of seakeeping requirements and their level of fulfillment on an actual ship.

Keywords: coastal patrol ship (CPS); full-scale seakeeping trials; ship design

---

1. Introduction

1.1. Literature Review

Full-scale ship sea trials are the most important test of a ship's structure, equipment, and crew in order to prove its security, reliability, and operational capability. In principle, a ship's trials should provide the final check of the adequacy of theoretical and experimental predictions of ship behavior [1]. Ship trials are carried out for a variety of reasons, including to:

1. Confirm that the ship meets her design intention as regards performance;
2. Predict performance during service;
3. Prove that equipment can function properly in the shipboard environment;
4. Provide data on which future ship designs can be based; and
5. Determine the effect on human performance.

Ship seakeeping design predictions are based on numerical analyses and model tests that are conducted in model tanks. Validation, using full-scale experimental data, is essential to the development of a ship's motion prediction code [2]. Reference [2] presents validations of ship motion predictions using model tank tests and full-scale sea trials for a Canadian naval destroyer. Full-scale trials are considered crucial because they include physical phenomena lacking in model tests due to oversight or scaling effects.

The results of full-scale seakeeping trials on several ships, including the Dutch naval destroyer "HM Groningen" are described in [3]. The main goal of these tests was to compare full-scale test results with numerical predictions that had been developed. It was noted that suitable conditions at sea are hard to find, or that the measuring or analyzing methods were not sufficient for correlation purposes.

A new experimental methodology to accurately predict wave-induced motions and load responses of ships was proposed in [4]. It is based on self-propelled large-scale model measurements that were conducted in natural environmental conditions. Onboard systems, operated by the crew, were used to measure and record sea waves and the responses of a model. A post-voyage analysis of the measurements, both of the sea waves and the model's responses, was conducted to predict the ship's motion and short term load responses to a corresponding sea state.

The results of extensive full-scale seakeeping trials of an all-weather lifeboat, conducted by the Royal National Lifeboat Institution in collaboration with Newcastle University and Lloyd's Register, are shown in Reference [5]. The trials investigated the seakeeping behavior of the craft in real operational conditions.

Patrol vessels must have good performance criteria for seaworthiness, and an analysis of the hydrodynamic aspects of a ship's design is one of the designer's primary tasks [6]. This analysis describes ship motion and ship resistance. The seakeeping tests conducted two variations of a ship loading condition and involved two sea states, namely the World Meteorological Organization (WMO) sea states 3 and 4.

In [7], the stern boat deployment system was investigated to evaluate the capability of launching and recovering a rigid hull inflatable boat (RHIB) via the stern ramp. The seakeeping characteristics required for the successful operation of a mother ship and inflatable boat were analyzed.

1.2. Paper Content and Main Particulars of the Coastal Partrol Ship

This paper presents the seakeeping requirements for a coastal patrol ship (CPS) for the Croatian Navy (CN) and the results of the full-scale seakeeping performance of the ship. The seakeeping requirements are based on Reference [8], which presents the seakeeping criteria for the reference ship, a frigate-size vessel. Because the CPS is a significantly smaller ship, the requirements are downsized accordingly, and presented in Section 2. The seakeeping requirements are composed of: the motion criteria of the ship (roll, pitch); vertical and lateral accelerations at significant positions on the ship; propeller emergence; deck wetness; slamming; and relative vertical motions of the edge of the ship stern. Full-scale trials were conducted during the acceptance of the ship, and the corresponding above-mentioned criteria were applied. After the ship was accepted by the CN, the prototype sea trials were conducted. The purpose of these tests was to check the utmost capabilities of the ship on higher sea states and see how those conditions affect the ship's hull, machinery, other equipment, and the crew. The results of full-scale trials are presented in Section 4. The main CPS particulars are shown in Table 1.

The seakeeping numerical prediction and model tests were done during preliminary project design, but these results are not comparable with the full-scale tests since the ship was significantly changed in the last phases of the project (e.g., lengthened).

The general arrangement and body plan of the CPS are shown in Figures 1 and 2. The main tasks for the new coastal patrol ship are to conduct:

1. Peacetime tasks:

   - Low enforcement at sea;
   - Protection of fishing;
   - Control and prevention of possible ecology incidents;
   - Combat against terrorism;
   - Trafficking of people and narcotics;

2. Miscellaneous tasks such as search and rescue and support of the local population in crises;

3. Tasks during wartime [9].

Table 1. The main coastal patrol ship (CPS) particulars.

Item	Specification
Length overall	43.5 m
Breadth overall	8.0 m
Breadth of hull at waterline	7.5 m
Draft over propellers	2.9 m
Maximum continuous speed	28 kn
Economy speed	15 kn
Block coefficient	0.45
Type of ship form	semi-displacement
Hull material	high strength steel
Superstructure material	Al alloy
Ship range	1000 nm
Armament	30 mm Aselsan bow gun 2x Browning machine guns 12.9 mm MANPADs
Propulsion	2x Caterpillar 16V, 3516C, 2525 kW
Main equipment	RHIB LoA 7.5 m

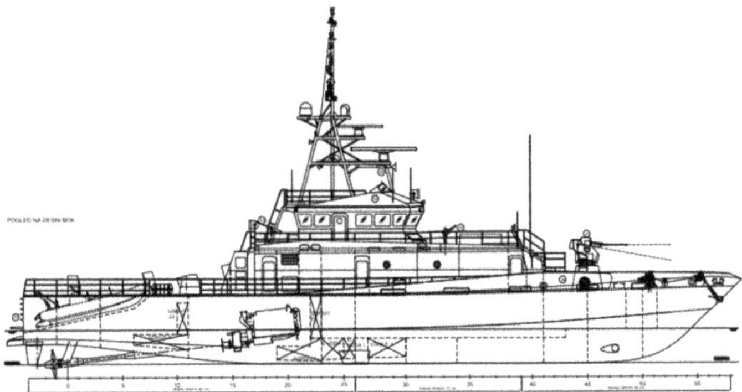

Figure 1. General arrangement of the CPS.

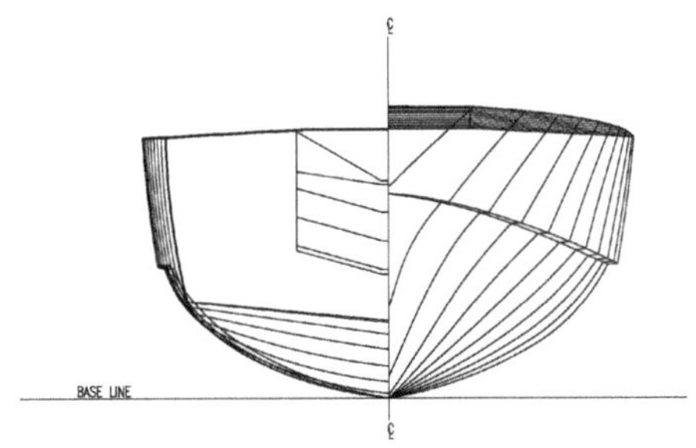

Figure 2. Body plan of the CPS.

2. Seakeeping Requirements

Seakeeping requirements are based on Reference [8], and are reduced to suit the CPS size. Following Reference [9], the CPS should be fully operational at mid sea state 4 ($H_{1/3}$ = 1.8 m) following the WMO scale, but not including the launch and recovery of the RHIB. Launch and recovery of the RHIB should be feasible at sea state 3 ($H_{1/3}$ = 1.25 m–WMO) and with the CPS sailing at least 5 knots with heading waves. The CPS should be partially operational up to sea state 5 ($H_{1/3}$ = 3.1 m–WMO) including sailing in a manner suitable for surveillance, monitoring the operational situation, and reporting. The requirements also include survivability of the CPS on the highest sea state observed on the Adriatic Sea. Other specific seakeeping requirements are shown in Table 2.

Table 2. Specific seakeeping requirements for the CPS.

	Requirement	Criteria
1.	Rolling	<4° RMS at sea state 4 ($H_{1/3}$ = 1.8 m) and speed of 15 knots, at sea state 3 ($H_{1/3}$ = 1.25 m) and speed of 5 knots, as well as in conditions of the sea trials, and at the maximum continuous speed.
2.	Rolling	<5° RMS at sea state 3 ($H_{1/3}$ = 1.25 m) speed of 0 knots, and beam waves.
3.	Pitching	<1.5° RMS at sea state 4 ($H_{1/3}$ = 1.8 m) and speed of 15 knots, at sea state 3 ($H_{1/3}$ = 1.25 m), and speed of 5 knots.
4.	Vertical accelerations at accommodation spaces	<0.2 g at sea state 4 ($H_{1/3}$ = 1.8 m) and speed of 15 knots, at sea state 3 ($H_{1/3}$ = 1.25 m) and speed of 5 knots, as well as in conditions of the sea trials, and at the maximum continuous speed.
5.	Lateral accelerations at accommodation spaces	<0.1 g at sea state 4 ($H_{1/3}$ = 1.8 m) and speed of 15 knots, at sea state 3 ($H_{1/3}$ = 1.25 m) and speed of 5 knots, as well as in conditions of the sea trials, and at the maximum continuous speed.
6.	Vertical displacement at the transom	<0.78 m at sea state 3 ($H_{1/3}$ = 1.25 m) and speed of 5 knots.
7.	Deck wetness	<30/h at sea state 4 ($H_{1/3}$ = 1.8 m) and speed of 15 knots.
8.	Slamming	<20/h at sea state 4 ($H_{1/3}$ = 1.8 m) and speed of 15 knots.
9.	Propeller emergence	<90/h at sea state 4 ($H_{1/3}$ = 1.8 m) and speed of 15 knots.

3. Test Conditions and Measuring Equipment

The trials were conducted with half-full fuel and water tanks, in addition to stocked supplies, at departure. A record sheet captured additional information: sea state; wind speed; wind direction; ship's heading; GPS coordinates; engine RPM; speed over ground; speed through water; draft and displacement of ship at departure; and trial times and duration. The sea states were estimated by three crew members with extensive nautical experience. They were based on sea state descriptions of the Adriatic Sea set by Prof. Tonko Tabain according to experiments conducted during the 1970s [10]. Prof. Tabain also set the relationship between the Adriatic sea state scale and the WMO scale. Sea state estimates were also compared with the sea states provided by the Croatian Meteorological Institute (CMI) to validate the estimates provided by the crew members. The CMI uses a network of buoys for wave measurement and providing related waves statistics. It was difficult to find the required sea state conditions and the trials therefore lasted longer than expected. To record parameters for all seakeeping criteria described in Section 2, the CPS sailed for 20 minutes for each set of criteria. For measuring rolling and pitching angles, a fibro optic gyro (FOG) onboard the CPS was used. The FOG provided raw motion data that were recorded on a laptop and processed later. A typical sequence of maritime courses for seakeeping tests was used as shown in Figure 3. Following Reference [11], root mean square (rms) values of signals were calculated using the expression:

$$\sqrt{(1/N)\sum_{i=1}^{N}\left(\delta-\bar{\delta}\right)^{2}} \tag{1}$$

where N represents the number of samples, δ is the measured signal value, and $\bar{\delta}$ is the mean value of the signal. The highest expected measured frequency was less than 0.5 Hz and the frequency of FOG data was 4.0 Hz, which assured consistent and well-sampled data. Accelerometers were used to measure vertical and lateral accelerations of specific points on the ship (i.e., the wheelhouse, the control cabin near the engine room, and the stern of the ship). Vertical displacement of the stern was measured using a liquid level sensor that was attached to the stern of the ship. Deck wetness, slamming, and propeller emergence were determined by counting their occurrences during one sailing course of 20 min, and then recalculating at the number of occurrences per hour.

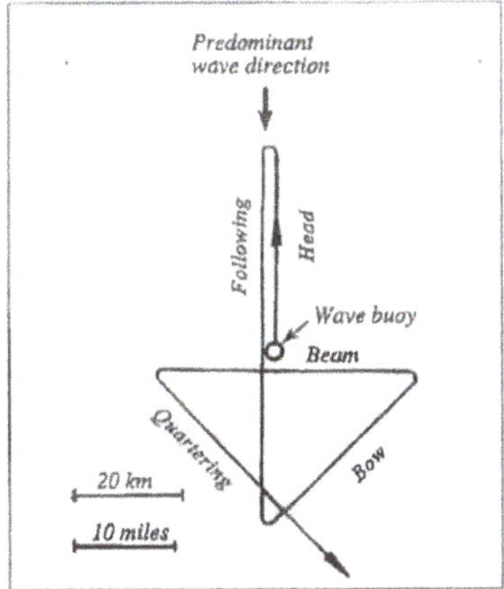

Figure 3. Sequence of maritime courses for seakeeping tests [12].

4. Seakeeping Performance Results

Seakeeping performance results are presented in Tables 3–5, and were taken from the report by the company hired by the Croatian Ministry of Defence (MoD) to conduct the trials [13].

Table 3 presents the results of the trial at sea state 3–4 (WMO), and at speeds of 0 and 5 knots. The speed of 5 knots was one of the required operational speeds designed for launch and recovery of the RHIB. One of the main goals, in addition to measuring all seakeeping parameters, was to check vertical displacement at the aft perpendicular (AP). This parameter is important for assessing the safe launch and recovery of the RHIB. For the operational sea state 3–4 (WMO), this parameter was 0.6 m, a result below the required limits and signaling the success of this crucial operation. This was born out in practice, with the RHIB successfully being launched and recovered during sea trials. Later on, when the crew had become more experienced, they launched and recovered the RHIB even at sea state 4–5 (WMO). All other parameters were within the required limits.

Table 4 presents the results of the trial at sea state 3–4 (WMO), and at a speed of 15 knots, representing the cruising speed of the CPS. As can be seen, all parameters were within required limits, except the angle of pitch (1.9° rms) at the heading angle of 90°, and the number of slammings that were slightly above the allowable limits.

Table 5 shows the results of the trial at sea state 3–4 (WMO), and at a speed of 27 knots (i.e., maximum continuous speed). The parameters that exceeded the required thresholds were the angle of pitching at the heading of 45° and 135°, deck wetness (33), and slamming (70). It was expected that the parameters for maximum continuous speed would be exceeded, although there were no requirements to measure these parameters at this speed.

Table 3. Results for the trial at sea state 3–4 (WMO), and speed of 0 and 5 knots.

Parameter	Symbol	Units	Trial I	Trial II	Trial III	Trial IV	Trial V	Trial VI
Initial ship speed	V_0	kn	0	5	5	5	5	5
Ship draft at AP	TKA	m	2.0	2.0	2.0	2.0	2.0	2.0
Ship draft at FP	TKF	m	1.98	1.98	1.98	1.98	1.98	1.98
Ship heading	COG	°	030	205	075	030	345	300
Sea state	WMO	-	3–4	3–4	3–4	3–4	3–4	3–4
Waves heading	-	°	90	0	45	90	135	180
Average speed (SOG)	v	kn	1.7	5.3	4.7	4.5	4.8	5.1
Roll	Φ	°	2.5	1.7	3.3	2.3	2.7	2.6
Pitch	θ	°	0.7	0.9	0.7	0.7	0.6	1.1
Vertical accel. at wheelhouse	w'	m/s^2	0.3	0.3	0.3	0.3	0.3	0.2
Lateral accel. at wheelhouse	v'	m/s^2	0.5	0.7	0.6	0.7	0.7	0.7
Vertical displ. at the transom	h_{2Pk-Pk}	m	N/A	0.60	N/A	N/A	N/A	N/A

Table 4. Results for the trial at sea state 3–4 (WMO), at 15 knots.

Parameter	Symbol	Units	Trial I	Trial II	Trial III	Trial IV	Trial V
Initial ship speed	V_0	kn	15	15	15	15	15
Ship draft at AP	TKA	m	2.0	2.0	2.0	2.0	2.0
Ship draft at FP	TKF	m	1.98	1.98	1.98	1.98	1.98
Ship heading	COG	°	205	130	270	310	345
Sea state	WMO		3–4	3–4	3–4	3–4	3–4
Waves heading	-	°	0	45	90	135	180
Average speed (SOG)	v	kn	14.9	15.5	15.9	15.8	15.7
Roll	Φ	°	1.0	2.2	2.7	2.7	1.8
Pitch	θ	°	0.9	1.5	1.9	1.3	0.8
Vertical accel. at wheelhouse	w'	m/s^2	0.4	0.9	0.5	0.4	0.3
Lateral accel. at wheelhouse	v'	m/s^2	0.2	0.6	0.7	0.6	0.4
Deck wetness	-	No. of occur./h	0	0	0	0	0
Slamming	-	No. of occur./h	24	20	0	0	0
Propeller emergence	-	No. of occur./h	0	0	0	0	0

Table 5. Results for the trial at sea state 3–4 (WMO), and speed of 27 knots.

Parameter	Symbol	Units	Trial I	Trial II	Trial III	Trial IV	Trial V
Initial ship speed	V_0	kn	27	27	27	27	27
Ship draft at AP	TKA	m	2.0	2.0	2.0	2.0	2.0
Ship draft at FP	TKF	m	1.98	1.98	1.98	1.98	1.98
Heading	COG	°	165	125	085	310	345
Sea state	WMO	-	3–4	3–4	3–4	3–4	3–4
Waves heading	-	°	0	45	90	135	180
Average speed (SOG)	v	kn	26.7	26.9	26.7	27.3	26.0
Roll	Φ	°	2.0	2.6	3.2	2.9	3.6
Pitch	θ	°	2.5	2.2	1.5	2.2	2.1
Vertical accel. (at wheelhouse)	w′	m/s^2	1.3	1.2	0.7	0.4	0.5
Lateral accel. (at wheelhouse)	v′	m/s^2	0.6	0.7	0.7	0.6	0.7
Deck wetness	-	No. of occur./h	33	0	0	0	0
Slamming	-	No. of occur./h	70	0	0	0	0
Propeller emergence	-	No. of occur./h	0	0	0	0	0

Tables 6 and 7 show the results of numerical calculations, based on strip theory and Jonswap spectra, conducted during the preliminary phase of ship design.

Measuring signal examples for the angles of pitching and rolling at v = 15 kn and wave heading = 45° are given on Figures 4 and 5 respectively.

In further trials, the prototype of the vessel was tested at the higher sea states 4–5 (WMO), and the results obtained were more than satisfactory. Even in these sea state conditions, the RHIB could be successfully launched and recovered, and there was negligible loss of speed. Generally, the vessel demonstrated very good seakeeping characteristics.

Table 6. Results of numerical calculations for sea state 3–4 (WMO), and ship speed of 5 knots.

Parameter	Symbol	Units	Trial I	Trial II	Trial III	Trial IV	Trial V	Trial VI	Trial VII
Waves heading			0°	30°	60°	90°	120°	150°	180°
Average speed	v	kn	5.0	5.0	5.0	5.0	5.0	5.0	5.0
Roll	Φ_{RMS}	°	0.0	3.52	7.15	9.01	12.60	4.59	0.0
Pitch	θ	°	1.21	1.29	1.31	0.57	1.18	1.28	1.2
Vertical accel. (at wheelhouse)	w′	m/s^2	0.348	0.435	0.675	0.788	0.451	0.106	0.063
Lateral accel. (at wheelhouse)	v′	m/s^2	0.0	0.866	1.633	2.004	2.609	0.508	0.0
Vertical displ. at the transom	$h_{2Pk\text{-}Pk}$	m	0.406	N/A	N/A	N/A	N/A	N/A	N/A

Table 7. Results of numerical calculation for sea state 3–4 (WMO), and ship speed of 15 knots.

Parameter	Symbol	Units	Trial I	Trial II	Trial III	Trial IV	Trial V	Trial VI	Trial VII
Waves heading			0°	30°	60°	90°	120°	150°	180°
Average speed	v	kn	15.0	15.0	15.0	15.0	15.0	15.0	15.0
Roll	Φ_{RMS}	°	0.0	2.6	6.99	10.41	6.5	3.37	0.0
Pitch	θ	°	1.73	1.72	1.53	0.64	1.01	1.07	1.05
Vertical accel. (at wheelhouse)	w′	m/s²	1.360	1.362	1.291	0.83	0.205	0.15	0.232
Lateral accel. (at wheelhouse)	v′	m/s²	0.0	0.786	1.636	2.225	0.934	0.609	0.0

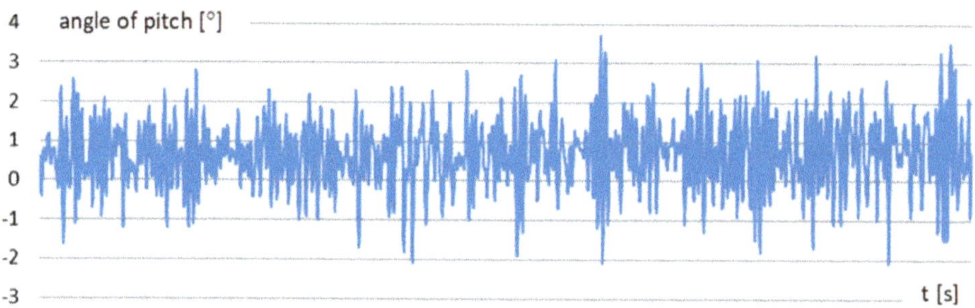

Figure 4. Measuring signal example—the angle of pitching at v = 15 kn and wave heading = 45°.

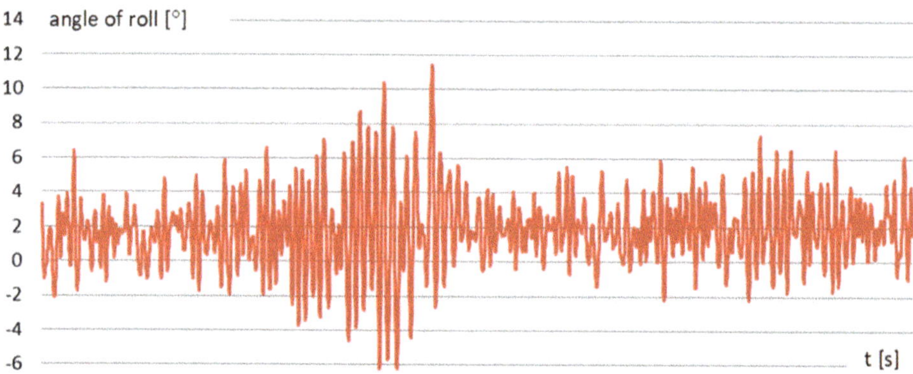

Figure 5. Measuring signal example—the angle of rolling at v = 15 kn and wave heading = 45°.

5. Discussion

The presented seakeeping results and performances of the CPS show very good seaworthiness capabilities of the ship. The ship is fully operational at the mid sea state 4 (WMO), which was one of the main seakeeping criteria. When the crew had enough experience manipulating the RHIB, the RHIB could be launched and recovered even at the upper limit of sea state 4. During tests on higher sea states 5 (WMO), the ship showed reasonable seakeeping characteristics, a high level of reliability of all ship subsystems, and a low level of speed loss on waves. Considering some specific criteria that were exceeded (e.g., slamming at maximum continuous speed in a head sea), this was expected and does not mean that ship has unsatisfactory seakeeping performance. There were also some deviations in the angle of pitching at a wave direction of 0 and 45 degrees that were unsurprising due to the size of the ship.

Upon increasing ship speed from 5 to 15 knots (cruising speed), the seakeeping parameters changed as follows:

- Roll angle became lower for the wave headings 0° and 180° due to the hydrodynamic stabilization of ship, but it became higher for beam seas (90°).
- Pitch angle was generally higher, except for a wave heading of 0°.
- Vertical accelerations were higher for all wave headings.
- Lateral accelerations had a similar pattern to the rolling of the ship.

Upon raising the ship speed to maximum continuous, it can be observed that all seakeeping parameters rose significantly, and the highest rising rate was related to the vertical acceleration, slamming, and deck wetness, which was expected especially from the heading seas.

Comparing numerical calculations and full-scale test results, the following conclusions can be drawn:

- Numerical calculations showed much higher roll and lateral acceleration responses because the numerical model could not take into account the influence of viscous forces on ship roll by ship appendages such as bilge keels, stern trim plate, etc.
- Numerical prediction of pitch angle and vertical accelerations showed reasonable agreement.

Seakeeping is one of the most important characteristics of a ship, and there is a need to continuously and rigorously consider it throughout all phases of ship design. This was the case with the CPS; seakeeping was tested through numerical analyses, model tests, and finally, full-scale tests on the ship. Setting seakeeping requirements at an early phase of the project is crucial for successful ship design, and the results presented herein indicate that it is reasonable and useful to set these seakeeping requirements for patrol ships of the CPS' size, as it helps designers to satisfy them.

Author Contributions: A.L. contributed to the paper by establishing the concept of paper and definitions of research goals and aims. He participated in conducting the research, control of experiments, and data collection and integration. He wrote an initial paper draft including data and graphics design and presentation. V.S. contributed to the paper through its review, validation, and its final preparation for publication. He participated in the analysis of references dealing with this research area. He also did management and coordination of the research planning. He is responsible for the presentation and interpretation of the work. All authors have read and agree to the published version of the manuscript.

Funding: This research received no external funding.

Conflicts of Interest: The authors declare no conflict of interest.

Nomenclature

AP	Aft perpendicular
CMI	Croatian Meteorological Institute
COG	Course over ground
CPS	Coastal Patrol Ship
FOG	Fiber Optic Gyro
FP	Fore perpendicular
g	Acceleration of gravity (9.80665 m/s^2)
GPS	Geographical Positioning System
$H_{1/3}$	Significant Wave Height
MANPAD	Man Portable Air Defense
MoD	Ministry of Defense
N	The number of measured samples
NATO	North Atlantic Treaty Organization
RHIB	Rigid Hull Inflatable Boat
RMS	Root Mean Square
RPM	Rotation per Minute
SOG	Speed over ground

STANAG	Standardization Agreement
δ	Measured signal value
$\bar{\delta}$	Average of measured signal value

References

1. Molland, A.F. *The Maritime Engineering Reference Book*; Elsevier Ltd.: Oxford, UK, 2008.
2. Kevin McTaggart, D.S. Validation of Ship Motion predictions with Sea Trials data for a Naval Destroyer in Multideirectional Seas. In Proceedings of the 25th Symposium on Naval Hydrodynamics, St. Johns, NL, Canada, 8–13 August 2004; pp. 1–13.
3. Gerritsma, J. Results of Recent Full Scale Seakeeping Trials. *Int. Shipbuild. Prog.* **1980**, *27*, 278–289. [CrossRef]
4. Jiao, J.; Sun, S.; Ren, H. Predictions of Wave Induced Ship Motions and Loads by Large-Scale Model Measurement at Sea and Numerical Analysis. *Brodogradnja* **2016**, *67*, 82–100. [CrossRef]
5. Prini, F.; Benson, S.; Birmingham, R.W.; Dow, R.S.; Ferguson, L.; Sheppard, P.J.; Phillips, H.J.; Johnson, M.C.; Mediavilla, J.; Hirdaris, S.E. Full-Scale Seakeeping Trials of an All-Weather Lifeboat. In Proceedings of the Surveillance, Search and Rescue Craft, SURV 9, London, UK, 18 April 2018.
6. Firadus, N.; Baharuddin, A. Hydrodynamic Analysis of Patrol Vessel Based on Seakeeping and Resistance Performance. *J. Ocean Mech. Aerosp.* **2018**, *60*, 1–6.
7. Ho Hwan Chun, E.A. Experimental Investigation on Stern-Boat Deployment System and Operability for Korean Coast Guard Ship. *Int. J. Nav. Archit. Ocean Eng.* **2012**. [CrossRef]
8. Military Agency for Standardisation (MAS). *Common Procedure for Seakeeping in the Ship Design Process*; NATO Standardization Agency: Brusells, Belgium, 2000.
9. Brodarski Institute. *Technical Specification of Coastal Patrol Ship*; Brodarski Institute: Zagreb, Croatia, 2014.
10. Tabain, T. Standard Wind Wave Spectrum for the Adriatic Sea Revisited (1977–1997). *Brodogradnja* **1997**, *45*, 303–313.
11. Jasna Prpić-Oršić, V.Č. *Pomorstvenost Plovnih Objekata*; Zigo: Rijeka, Croatia, 2006.
12. Lloyd, A. *SEAKEEPING: Ship Behaviour in Rough Weather*; A R J M Lloyd: Hampshire, UK, 1998.
13. BoBLab Ltd. *Report of Seakeeping Tests for Coastal Patrol Ship OOB-31*; BobLab Ltd.: Zagreb, Croatia, 2018.

Article

A Theoretical Study on the Hydrodynamics of a Zero-Pressurized Air-Cushion-Assisted Barge Platform

Fengmei Jing [1], Li Xu [2], Zhiqun Guo [3,*] and Hengxu Liu [3]

[1] School of Mechanical Engineering, Beijing Institute of Petrochemical Technology, Beijing 102617, China; Jingfengmei@bipt.edu.cn
[2] Shanghai Branch, China Ship Scientific Research Center, Shanghai 200001, China; xu_li@hrbeu.edu.cn
[3] College of Shipbuilding Engineering, Harbin Engineering University, Harbin 150001, China; liuhengxu@hrbeu.edu.cn
* Correspondence: guozhiqun@hrbeu.edu.cn; Tel.: +86-451-8258-9204

Received: 8 July 2020; Accepted: 22 August 2020; Published: 27 August 2020

Abstract: Thebarge platform has the advantages of low cost, simple structure, and reliable hydrodynamic performance. In order to further improve the hydrodynamics of the barge platform and to reduce its motion response in waves, a zero-pressurized air cushion is incorporated into the platform in this paper. The pressure of the zero-pressurized air cushion is equal to atmospheric pressure and thus does not provide buoyancy to the platform. As compared to the conventional pressurized air cushion, the zero-pressurized one has advantages of less air leakage risk. However, due to the coupling effect on the interface between water and air cushion, the influence of the gas inside the air cushion on the performance of the floating body has become a difficult problem. Based on the boundary element method, the motion response of the zero-pressurized air-cushion-assisted barge platform under regular and irregular waves is calculated and analyzed in the paper. Compared with the barge platform without air cushion, numerical results from the theoretical method show that in regular waves, the air cushion could significantly reduce the amplitude of heave and pitch (roll) response of the round barge platform in the vicinity of resonance. In irregular waves, the air cushion also observably reduces the pitch (roll) motion, though amplifies the heave motion due to the transfer of heave resonance frequency. Thetheoretical study demonstrates that the zero-pressurized air cushion can reduce the seakeeping motion of barge platforms in high sea states, but might also bring negative effects to heave motion in low sea states. One should carefully design the air cushion for barge platforms according to the operating sea states to achieve satisfactory hydrodynamic performance in engineering application.

Keywords: barge platform; zero-pressurized air cushion; hydrodynamic performance; boundary element method

---

1. Introduction

The round barge is a common floating platform in ocean engineering. Its structure is simple, the cost is cheaper than other floating platforms, and its life is longer (about 100 years). However, due to the large waterline area of the barge, the motion response under the incident waves is also relatively large. How to improve the seakeeping performance of a barge and reduce its motion response in waves has always been the focus of research in ocean engineering.

At present, anti-roll tanks (ARTs), tunedliquid column dampers (TLCDs) and tuned mass dampers (TMDs), air cushions (ACs), heave bottom plates (HBPs), and so forth, can be used to reduce the motion of the offshore floating barge [1]. The principle of the tuned water column damper is the same as the anti-roll water tank, which are considered as one kind in this paper. It was reported that, as compared to other dampers, the air cushion has the most significant anti-rolling effect on the floating platform [1]. Results of studies in other literature also confirm that the air cushion has an obvious effect on improving the hydrodynamic performance of the barge-type offshore platform. The air cushion can significantly reduce the wave bending moment of the floating platform, which is because the air cushion disperses the relatively concentrated wave load [2], and chronic drift forcedue to the existence of the free surface in the air cushion makes waves relatively easy to pass through the platform [3,4].

The application of air cushion technology in the field of ocean engineering has a long history, which can be traced back to the 1970s [3]. Pinkster and Fauzi studied a square air-cushion-supported structure [5]. The Green function method and three-dimensional linear radiation/diffraction theory were used to predict its motion response under waves and the air pressure inside the air cushion, and the numerical results agree well with the experimental ones [6]. Ikoma et al. [7] had studied the elastic floating platform supported by the air cushion based on the potential flow theory and pressure distribution method. The integral equation was used to calculate the movement of the internal air cushion under the regular waves. The results show that the air cushion can effectively reduce the wave drift force and the motion response of the floating platform [7]. For the verylarge floating body supported by the air cushion, Kessel analyzed its motion response based on the three-dimensional linear potential flow theory. The results suggest that this method can accurately solve the motion response of the floating body. It has been improved, and the wave bending moment acting on the structure has been significantly reduced [8]. Lee and Newman studied the wave effects of a super-large floating air cushion support platform in waves based on potential flow theory. They used a series of given Fourier modes to represent the vertical motion on the free surface in the air cushion of the air-supported floating structure, and thus extended the traditional six-freedom rigid body motion equation. The effect of air movement inside the air cushion is expressed by the derived aerodynamic added mass coefficient [9]. Lee and Newman made further improvements using the generated Fourier modals to represent changes in the internal oscillation pressure of the air cushion, which had been applied in WAMIT software [10]. Bie et al. [11] pointed out that the stability of the air-cushion-supported structure was lower than that of the common floating body under the same conditions, while the air cushion compartmentalization can improve the stability. Zhang et al. [12] conducted experimental and theoretical studies on the floating stability of an air-cushion-supported artificial island, whose foundation consists of multiple air cylinder structures.

On the other hand, the air cushion has also been applied to the ship field. Yang et al. [13] numerically and experimentally studied the seakeeping performance of a partial air-cushion-supported catamaran (PACSCAT) sailing in regular waves. Yang et al. [14] and Cucinotta et al. [15–17] investigated the air cavity and its evolution under stepped planning hulls. However, the air cushion/cavity under ships generally involves physical processes such as air inflow/generation and air leakage, which are more complicated than thoseunder platforms. Actually, the air cushion under platforms is enclosed by platform structures and free surface, and usually is isolated from atmosphere, so its hydrodynamic performance can be analyzed using simpler theoretical models.

In summary, the air cushion possessesa certain amount of displacementin the above literature, which might undergo risk of air pressure loss.On the contrary, the floating platform fitted with a zero-pressurized air cushion has a relatively high safety performance. That is because its static pressure is equal to the atmospheric pressure, and there is no need to worry about the damage or leakage of the air tanks. However, the hydrodynamic performance of the zero-pressurizedaircushion platform is rarely studied.

A round barge platform with zero-pressurized air cushion is proposed in this paper, and its hydrodynamic performance is studiedbased on the boundary element method. Firstly, the boundary conditions and control equations are constructed, and the air velocity potential is solved to obtain the air cushion aerodynamic coefficient; secondly, the barge platform motion equations are established and solved, where the hydrodynamic coefficients and wave force are obtained; finally, the influence of the zero-pressurized air cushion on the hydrodynamic performance of the barge platform is studied and the effect of the water depth on the air-cushion-assisted barge platform is analyzed.

2. The Zero-Pressurized Air-Cushion-Assisted Barge Platform

The three-dimensional model of the zero-pressurized air-cushion-assisted barge platform is shown in Figure 1, and the design parameters are shown in Table 1.

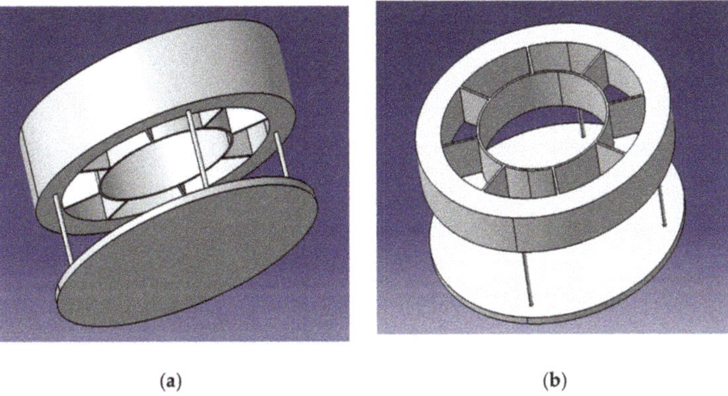

(a)   (b)

Figure 1. 3D model diagram of the zero-pressurized air-cushion-assisted barge platform. (**a**) Overall model diagram; (**b**) underwater structure model diagram.

Table 1. Parameters of the zero-pressurized air-cushion-assisted barge platform.

Parameter	Numerical Value	Unit
Design water depth (h)	≥30	m
Outer diameter of floating tank (d_1)	2 × 30 = 60	m
Outer diameter of air tank (d_2)	2 × 22.5 = 45	m
Diameter of middle tank (d_3)	2 × 5.8 = 11.6	m
Subdivision	Eighth Division	
Platform height	25	m
Platform draft (d)	20	m
Platform freeboard (h_f)	5	m
Platform mass	1251	t
Ballast material	Placer/Concrete	
Ballast diameter (d_4)	2 × 30 = 60	m
Ballast height	0.354	m
Ballast mass	2000	t

The zero-pressurized air-cushion-assisted bargeplatform consists of a barge structure main body at the top, a ballast plate at the bottom with the function of heave damping and ballast together, and a connecting truss that connects the above two parts. The main body of the bargeplatform includes a zero-pressurized air cushion tank and a buoyancy tank. The air cushion tank is divided into eightcompartments, which are geometrically equal and symmetrical sector-annular air tanks. These eightcompartments do not provide buoyancy, but they can act as an air spring when the bargeplatform heaves to reduce the motion response of the platform. The buoyancy tank on the periphery of the air tank is a structure that mainly provides buoyancy to the platform. Various materials can also be used to enhance the structural strength of the platform. The bottom of the buoyancy tank has a ring of damping skirt to increase the damping of the platform and thus reduce the motion response in waves. The truss can move up and down through the buoyancy tank to adjust the draft of the ballast tank, which is beneficial for towing the platform through shallow water. The detailed structure of the aircushion floating platform is shown in Figures 2 and 3.

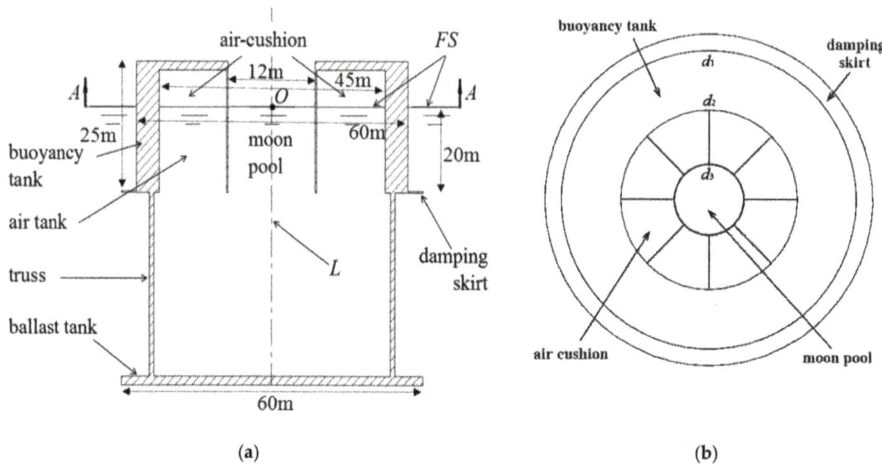

Figure 2. Schematic diagram of the zero-pressurized air-cushion-assisted barge platform. (**a**) Side view; (**b**) top view.

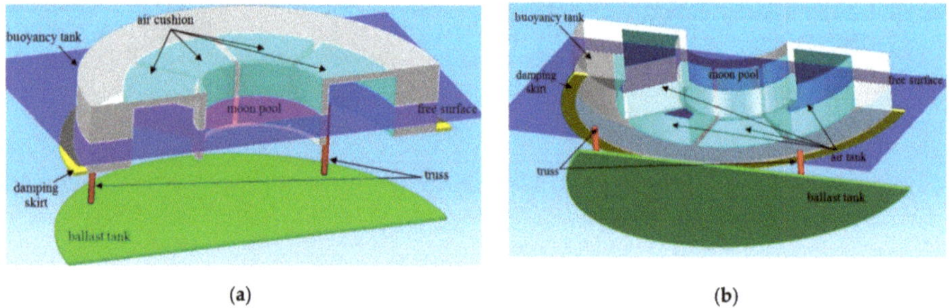

Figure 3. Schematic diagram in 3D of the zero-pressurized air-cushion-assisted barge platform. (**a**) Top side view; (**b**) bottom side view.

3. Motion Equations of the Air-Cushion-Assisted Barge Platform

Assuming that the fluid is ideal and the flow is irrotational, and the air in the cushion is compressible, the motion of the zero-pressurized air-cushion-assisted barge platform in the waves can be analyzed using potential flow theory. Let S_b be the wetted surface of the barge platform, S_c the inner surface of the platform that surrounds the air cushion, and S_i the air–water interface under the air cushion. Then, the complete enclosed surface surrounding the air cushion is expressed as $S_a = S_c + S_i$; the complete boundary between the water and the barge platform can be expressed as $S_w = S_b + S_i$. The free surface outside the float is denoted by S_f. The height of the air tank is h, and the draft is d.

3.1. The Definite Problem for the Air Cushion

Within the linear frequency domain conditions, the air velocity potential in the cushion can be presented as

$$\Psi = \mathrm{Re}\{\psi e^{i\omega t}\} \tag{1}$$

Obviously, the air in cushions should obey the law of mass, momentum, and energy conservation [18], from which one gets the control equation (Helmholtz equation) for ψ

$$\nabla^2\psi + K_a{}^2\psi = 0 \tag{2}$$

where $K_a = \frac{\omega}{c_0}$, and $c_0 = \sqrt{\frac{dp}{d\rho_a}}$ which is the acoustic velocity under adiabatic conditions, p is the atmospheric pressure, ρ_a is the air density.

Since the air cushion in the platform is fan-shaped (as shown in Figure 4), the cylindrical coordinate system is appropriate to describe the definite problem. Let a, b be the inner and outer diameter of the fan-shaped cushion, respectively. The angle between the boundary of one side and the starting coordinate axis in the positive direction is $\theta = c$, and the angle between the boundary of the other side and the starting coordinate axis in the positive direction is $\theta = d$, the height of air tank is $z_h - z_l = h$. The control equation (Helmholtz equation) satisfied by the air velocity potential in the fan-shaped air cushion in the cylindrical coordinate system can be written as

$$\frac{1}{r}\frac{\partial}{\partial r}\left(r\frac{\partial \psi}{\partial r}\right) + \frac{1}{r^2}\frac{\partial^2 \psi}{\partial \theta^2} + \frac{\partial^2 \psi}{\partial z^2} + K_a^2 = 0 \tag{3}$$

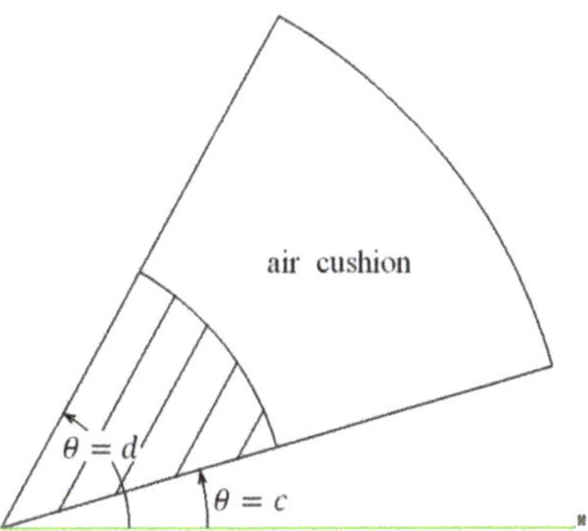

Figure 4. Top view of single fan-shaped air tank.

From the Bernoulli equation, the relationship between the air velocity potential ψ and the pressure p can be obtained. So the pressure on the free surface that is under the air cushion can be expressed as

$$p(x, y, z_l) = -\rho g \sum_{j=7}^{6+N_p} \xi_j n_j(x, y) \tag{4}$$

where ρ is water density.

The free surface condition for the air velocity potential in the air cushion is obtained:

$$\psi = \sum_{j=7}^{6+N_p} \frac{\rho g}{i\omega \rho_a} \xi_j n_j(x, y) \tag{5}$$

According to numerical tests in literature [10], if the air cushion is not too large, the uniform pressure can obtain satisfactory results. In this paper, the CFD (Computational Fluid Dynamics) simulation result also suggests that the variation of air pressure in a cushion is no more than 0.2%. Therefore, the pressure on the free surface can be assumed to be evenly distributed, that is, one can set $N_p = 1$, $n_7(x, y) = 1$.

The wall conditions for the air velocity potential in the cushion are:

$$\frac{\partial \psi}{\partial n} = i\omega \sum_{j=1}^{6} \xi_j N_j \tag{6}$$

Decomposing the air velocity potential yields

$$\psi = i\omega \sum_{j=1}^{6+N_p} \xi_j \Phi_j \tag{7}$$

Then the air velocity potential component Φ_j also satisfies the control Equation (3), and when $j \leq 6$, the following boundary conditions are obtained:

$$\begin{cases} \dfrac{\partial \Phi_j}{\partial n} = N_j & \text{on } S_c \\ \Phi_j = 0 & \text{on } S_i \end{cases} \tag{8}$$

When $j \geq 7$, there exist the following relations:

$$\begin{cases} \dfrac{\partial \Phi_j}{\partial n} = 0 & \text{on } S_c \\ \Phi_j = -\dfrac{\rho}{\rho_a}\dfrac{g}{\omega^2} n_j(x,y) & \text{on } S_i \end{cases} \tag{9}$$

Equations (3), (8), and (9) constitute the definite Helmholtz problem for the air velocity potential in the air cushion, which can be solved by analytical or numerical methods.

For the rectangular air cushion compartments, the WAMIT software [10] that released by the Massachusetts Institute of Technology can directly give the analytical solution for the air velocity potential [11]. However, for the fan-shaped air cushion, there exists neither analytical solver nor analytical solution given in literature that can directly solve the abovementioned definite problem. To this end, the open-source program BEMHELM [19] that released by the Nantes Central Institute of Technology is employed to numerically solve the definite Helmholtz problem.

After obtaining the velocity potential Φ_j, the air cushion pressure can be written as

$$P(x,y,z) = \rho_a \omega^2 \sum_{j=1}^{6+N_p} \xi_j \Phi_j \tag{10}$$

3.2. Motion Equations for Air-Cushion-Assisted Barge Platform

The air dynamic expression in frequency domain is:

$$f_a = \iint_{S_c} P(x,y,z) \cdot N_i dS = \sum_{j=1}^{6+N_p} \left(\omega^2 \mu_{ij}^a - i\omega \lambda_{ij}^a - C_{ij}^a \right) \xi_j \quad (1 \leq i \leq 6) \tag{11}$$

μ_{ij}^a, λ_{ij}^a, and C_{ij}^a are aerodynamic coefficients, where μ_{ij}^a is the air added mass matrix; λ_{ij}^a is the air damping coefficient matrix; C_{ij}^a is the air restoring coefficient matrix.

Combining Equation (11) with Equation (10), the following expression can be obtained:

$$\iint_{S_c} P(\xi) N_i dS = \rho_a \omega^2 \iint_{S_c} \left(\xi_j \Phi_j \right) N_i dS = \left(\omega^2 \mu_{ij}^a - i\omega \lambda_{ij}^a - C_{ij}^a \right) \xi_j \tag{12}$$

Based on the above section where the velocity potential Φ_j is solved, the aerodynamic coefficients can be obtained by the analytical or numerical methods.

Thus, the motion equations of the zero-pressurized air-cushion-assisted barge platform can be presented as

$$\sum_{j=1}^{6+N_p} \left(-\omega^2 \left(M_{ij} + \mu_{ij} + \mu_{ij}^a \right) + i\omega \left(\lambda_{ij} + \lambda_{ij}^a \right) + \left(C_{ij} + C_{ij}^a \right) \right) \xi_j = X_i \quad 1 \leq i \leq N_p \tag{13}$$

where M_{ij}, μ_{ij}, λ_{ij} represent the mass matrix, added mass matrix, and damping matrix, respectively, which can be solved by using the WAMIT software [10].

4. Motion Response of the Zero-Pressurized Air-Cushion-Assisted Barge Platform in Waves

The motion response of the zero-pressurized air-cushion-assisted barge platform is calculated by Equation (13), where the air dynamic and hydrodynamic coefficients are obtained using the BEMHELM [19] solver and WAMIT [10] software, respectively. The principal parameters of the zero-pressurized air-cushion-assisted platform are shown in Table 2. The panel model for the WAMIT calculation was established by Multisurf [10], as shown in Figure 5.

Table 2. Parameters of the barge platform with moon pool.

Parameter	Numerical Value	Unit
Design water depth (h)	$\geq$30	m
Outer diameter of floating tank (d_1)	$2 \times 30 = 60$	m
Inner diameter of floating tank (d_2)	$2 \times 22.5 = 45$	m
Platform height	25	m
Platform draft (d)	20	m
Platform freeboard (h_f)	5	m
Platform mass	1251	t
Ballast material	Placer/Concrete	
Ballast diameter (d_4)	$2 \times 30 = 60$	m
Ballast height	0.354	m
Ballast mass	2000	t

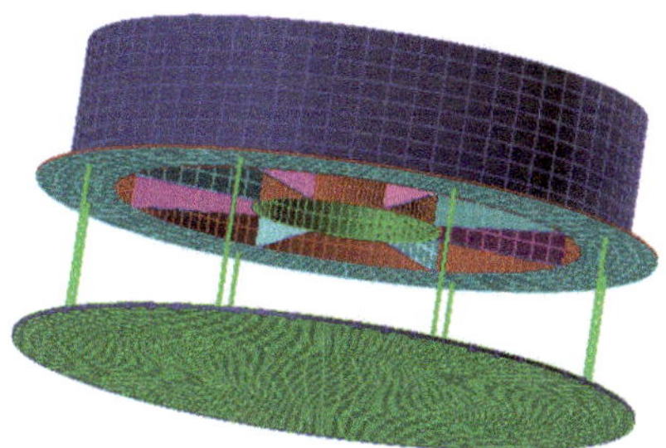

Figure 5. Panel model of the air-cushion-supported barge floating platform.

4.1. The Influence of the Zero-Pressurized Air Cushion on the Hydrodynamic Performance of the Barge Platform

In order to study the impact of the zero-pressurized air cushion on the hydrodynamic performance of the platform, a conventional barge platform (see Figure 6) is selected for comparison that is obtained by reducing the air cushion and its partition plates from the zero-pressurized air-cushion-assisted barge platform. The partition plate is a thin plate whose volume approximately equals to 0. To differentiate the two platforms, the zero-pressurized air-cushion-assisted barge platform is named as "Platform with air cushion", while the conventional barge platform is named as "Platform with moon pool".

The platform with moon pool has the same displacement as the one with aircushion. The water depth is set as 60 m, which is 3 times the platform draft. The response amplitude operators (RAOs) of the surge, heave, and pitch of the two platforms in regular waves are shown in Figure 7a–c, respectively, where the solid line denotes the results from the platform with air cushion, and the dashed lines represents those from the platform with moon pool.

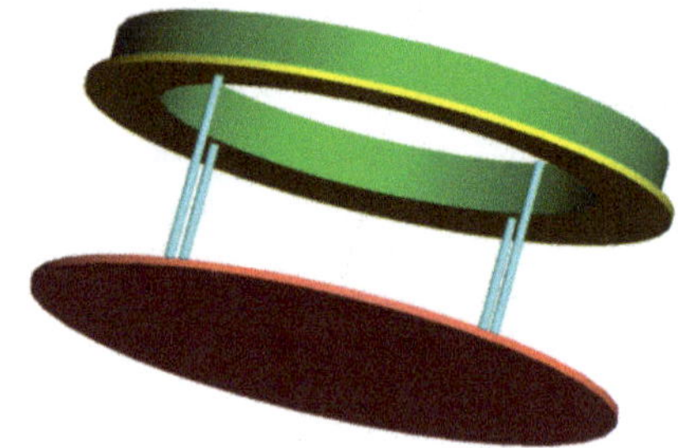

Figure 6. The platform model with moon pool.

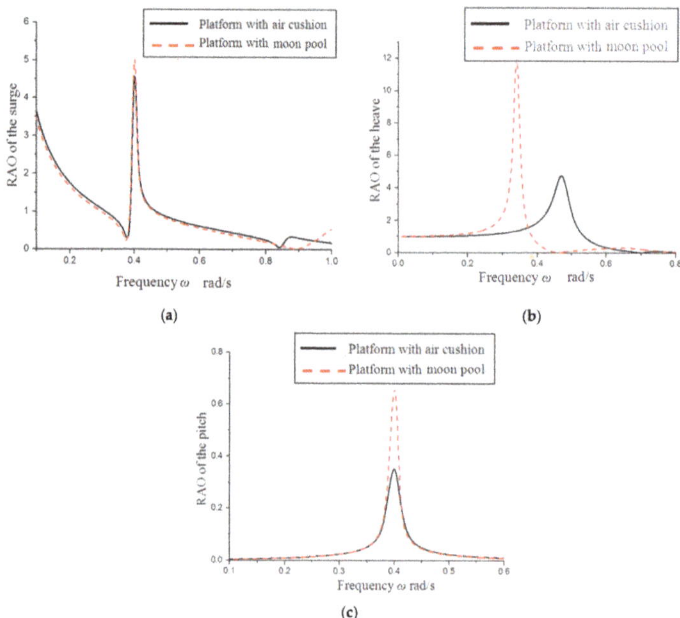

Figure 7. Comparison of RAOsbetween platform with air cushion andplatform with moon pool. (a) Surge RAO; (b) heave RAO; (c) pitch RAO.

As seen from Figure 7a, the curves of surge response from the two platforms almost coincide with each other. When the frequency is around 0.4 rad/s, the RAO reaches the maximum value, and the platforms resonate in the surge direction. The surge motion resonance range of the floating platform is narrow, which is between 0.4 rad/s and 0.6 rad/s. After the resonance area, the surge motion decreases rapidly with an increase of frequency. It can be found that the air cushion has little effect on the surge of the floating platform, which suggests that the air dynamics do not play an important role in the surge direction. This is because the surge motion does not change the volume of the air cushion.

As seen from Figure 7b, the air cushion has significant impact on the heave of the platform. The peak of the heave from the platform with air cushion is reduced about 60%, as compared with the big moon pool one. The heave motion can change the volume of the air cushion and thus excites significant air dynamics. It is worth noting that the resonance frequency of the platform is shifted by the air cushion from a lower frequency (from 0.33 rad/s to 0.36 rad/s) to a higher frequency (from 0.46 rad/s to 0.51 rad/s). The resonance frequency of the platform with air cushion is decided by the natural frequency of both barge platform and air cushion. In the engineering design, the resonance frequency of the platform with air cushion can be adjusted by changing the size of the air cushion to keep away from the classical wave frequencies.However, if the size of the air cushion is not tuned properly, the air cushion might increase the heave RAO of the barge platform at higher frequencies and have a negative effect on the barge platform response, as seen in Section 4.3.

As seen from Figure 7c, the pitch amplitude of the platform in the resonance interval can also be greatly reduced by the air cushion. The maximum reduction is about 50%. Obviously, in this case, the air dynamics are also excited by the change of air cushion volume due to the pitch motion. The difference between heave and pitch is that the pitch resonance frequency is not shifted by the air cushion. Except for the resonance interval, the air cushion has little impact on the pitch of the platform.

4.2. The Influence of Water Depth on the Hydrodynamic Performance of the Zero-Pressurized Air-Cushion-Assisted Barge Platform

The Chinese offshore water depth is around 30 m to 100 m, which is comparable to the draft of the barge platform, so the water depth might have significant influence on the hydrodynamic performance of the platform in the engineering application. To evaluate this effect, the RAO of the zero-pressurized air-cushion-assisted barge platform, with theshape/dimensions of the barge kept the same as in Section 4.1, are calculated in three water depths: 32.5 m, 60 m, and 100 m. The RAOs of the surge, heave, and pitch of the barge platform in the regular waves are given in Figure 8a–c, respectively, where the solid lines, dashed lines, and stippling lines represent the numerical results from 100 m, 60 m, and 32.5 m water depth, respectively.

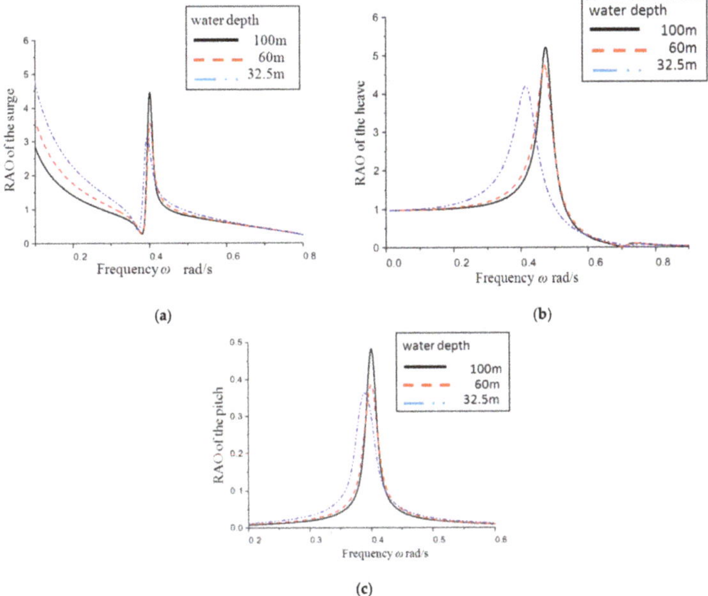

Figure 8. RAO of the platform with air cushion in different water depths. (a) Surge RAO; (b) heave RAO; (c) pitch RAO.

As seen from Figure 8a, the maximum surge RAO increases with the water depth, while the resonance frequency interval is barely affected by the water depth.

As seen from Figure 8b,c, the maximum heave (pitch) RAO increases with the water depth.Obviously, when the water depth is less than 60 m, the increasing water depth significantly affects the resonance frequency interval by shifting the resonance interval from lower to higher frequency.

In contrast, when the water depth is larger than 60 m, the resonance frequency interval is almost not affected by the increasing water depth, though the maximum heave (pitch) RAO increases with water depth.

In a word, when the water depth is less than 60 m, both the maximum RAO and the resonance frequency interval will be significantly affected by the water depth. However, with further increasing of the water depth, the influence on the resonance frequency interval can be ignored, while the maximum RAO still increases. The investigating results suggest that the zero-pressurized air-cushion-assisted barge platform has better performance in shallow water.

4.3. Motion of the Zero-Pressurized Air-Cushion-Assisted Barge Platform in Irregular Waves

To investigate the motion response of the zero-pressurized air-cushion-assisted barge platform in irregular waves, the Chinese offshore wave spectrum [20] was employed for simulating the real wave energy spectra, which is used to describe Chinese coastal waters. The expression of the China Sea spectrum is

$$S_\zeta(\omega) = \frac{A}{\omega^5} \exp\left(-\frac{B}{\omega^2}\right) \tag{14}$$

with

$$A = 0.74$$
$$B = \frac{g^2}{6.28^2 \cdot \overline{H}_{\frac{1}{3}}}$$

where $\overline{H}_{\frac{1}{3}}$ is the significant wave height.

Figure 9 portrays the significant value of surge, heave, and pitch response of the platformwith air cushion in irregular waves of sea states from 2 to 6.

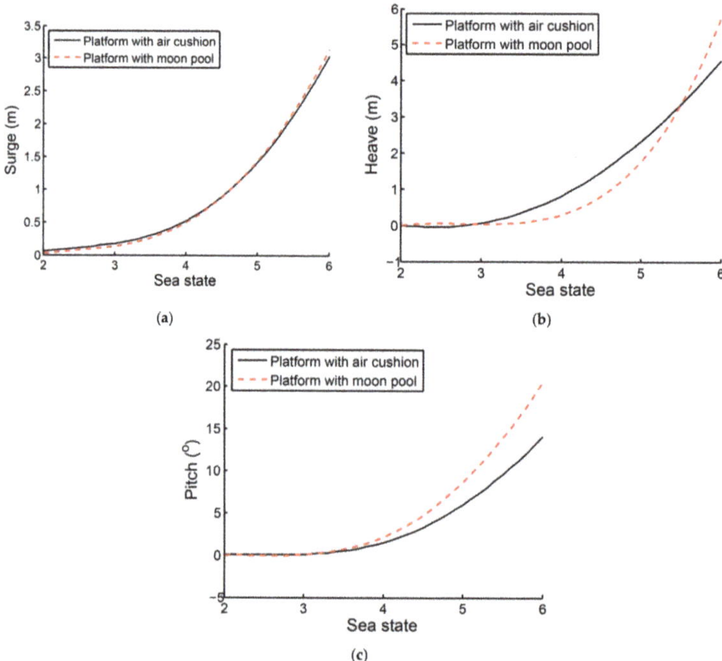

Figure 9. Significant value of motion response of the platforms in irregular waves. (**a**) Surge; (**b**) heave; (**c**) pitch.

From Figure 9a, one can observe that the surge response of the platform with air cushion is slightly larger than the moon pool one in low sea states (≤ 4), but smaller in high sea states (≥ 5). Similarly, in Figure 9b, the heave response of the platform with air cushion is larger than the moon pool one when sea state is no more than 5, but much smaller when sea state is beyond 6. Finally, from Figure 9c, one finds that the pitch response of the platform with air cushion is always smaller than the moon pool one, and the pitch reduction effect increases with the sea state.

Therefore, the air cushion can reduce the overall motions of the barge platform in high sea states, but might bring negative effects to heave motion in low sea states.

5. Conclusions

In this paper, the hydrodynamic performance of a zero-pressurized air-cushion-assisted barge platform was studied. The definite problem for the air velocity potential in the cushion was firstly proposed and theopen-source Helmholtz solver BEMHELM was employed to solve it. Then the motion equations of the zero-pressurized air-cushion-assisted barge platformwereestablished, in which the hydrodynamic coefficient and wave force of the barge platform were solved by the commercial hydrodynamic software WAMIT. Finally, the motion responses of the barge platform in regular and irregular waves were studied and the following conclusions were obtained.

(1) The zero-pressurized air cushion has a significant suppression effect on heave and pitch motion of the barge platform in the vicinity of the resonance frequencies (about 50% of the maximum motion response can be reduced).

(2) With the reduction of water depth (from 60 m to 30 m), the maximum motion RAO of the zero-pressurized air-cushion-assisted barge platform decreases, which suggests that the platform has excellent hydrodynamic performance in shallow water.

(3) The zero-pressurized air cushion can always reduce the pitch motion of the barge platform in irregular waves, as well as the heave motion in high sea states.However, the zero-pressurized air cushion might bring negative effects to heave motion in low sea states. One should carefully design the air cushion for barge platforms according to the operating sea states to achieve satisfactory hydrodynamic performance in engineering application.

Author Contributions: L.X. developed the theoretical method; Z.G. prepared the study cases; F.J. and H.L. performed the numerical calculation and analysis; and F.J. wrote the paper. All authors have read and agreed to the published version of the manuscript.

Funding: This research was funded by the National Natural Science Foundation of China (grant No.51779063, 51979065), State Key Laboratory of OceanEngineering (Shanghai Jiao Tong University) (Grant No.1913), and the Natural Science Foundation of Heilongjiang Province of China (Grant No. E2018025).

References

1. Borg, M.; Utrera Ortigado, E.; Collu, M.; Brennan, F.P. Passive damping systemsfor floating vertical axis wind turbines analysis. In Proceedings of the European Wind Energy Conference, EWEA, Vienna, Austria, 3–7 February 2013.

2. Pinkster, J.A.; Scholte, E.J.A.M. The behaviour of a large air-supported MOB at sea. *Mar. Struct.* **2001**, *14*, 163–179. [CrossRef]

3. Ikoma, T.; Masuda, K.; Rheem, C.K.; Maeda, H. Response Reduction of Motion and Steady Wave Drifting Forces of Floating Bodies Supported by Aircushions in Regular Waves: The 2nd Report—Response Characteristics in Oblique Waves. In Proceedings of the ASME 2007, International Conference on Offshore Mechanics and Arctic Engineering, San Diego, CA, USA, 10–15 June 2007; pp. 3–9.

4. Kurniawan, A.; Greaves, D.; Chaplin, J. Wave energy devices with compressible volumes. *Proc. R. Soc. A* **2014**, *470*, 20140559. [CrossRef] [PubMed]

5.   Pinkster, J.A.; Fauzi, A.; Inoue, Y.; Tabeta, S. The behaviour of large air cushion supported structures in waves. In Proceedings of the Second International Conference on Hydroelasticity in Marine Technology, Fukuoka, Japan, 1–3 December 1998; pp. 497–505.

6.   Tabeta, S. *Model Experiments on Barge Type Floating Structures Supported by Air Cushions*; Ship Hydromechanics Laboratory Report 1125; Delft University of Technology: Delft, The Netherlands, 1998.

7.   Ikoma, T.; Masuda, K.; Maeda, H.; Rheem, C. Hydroelastic Behavior of Air-Supported Flexible Floating Structures. In Proceedings of the 21st International Conference on OMAE'02, Oslo, Norway, 23–28 June 2002; Volume 2, pp. 745–752.

8.   Kessel, J.V. Air-Cushion Supported Mega-Floaters. Ph.D. Thesis, Delft University of Technology, Delft, The Netherlands, 2010.

9.   Lee, C.H.; Newman, J.N. Wave Effects on Large Floating Structures with Air Cushions. *Mar. Struct.* **2000**, *13*, 315–330. [CrossRef]

10.   Lee, C.H.; Newman, J.N. An Extended Boundary Integral Equation for Structures with Oscillatory Free-Surface Pressure. *Int. J. Offshore Polar Eng.* **2016**, *26*, 41–47. [CrossRef]

11.   Bie, S.A.; Shi, Z.M.; Wang, L.Y. Analysis of Floating Stability of Air Cushion Supported Structure at Small Inclination Angle. *China Harb. Constr.* **2001**, *2*, 31–35.

12.   Zhang, J.L. *Experimental Research on the Motion Performance of Air Supported Artificial Island Foundation Floating Towing*; Tianjin University: Tianjin, China, 2011.

13.   Yang, J.L.; Lin, Z.; Gao, Z.Y.; Li, P. A Study on the Motion of Partial Air Cushion Support Catamaran in Regular Head Waves. *Water* **2019**, *11*, 580. [CrossRef]

14.   Yang, D.M.; Sun, Z.Y.; Jiang, Y.; Gao, Z.Y. A Study on the Air Cavity under a Stepped Planing Hull. *J. Mar. Sci. Eng.* **2019**, *7*, 468. [CrossRef]

15.   Cucinotta, F.; Guglielmino, E.; Sfravara, F.; Strasser, C. Numerical and experimental investigation of a planing Air Cavity Ship and its air layer evolution. *Ocean Eng.* **2018**, *152*, 130–144. [CrossRef]

16.   Cucinotta, F.; Guglielmino, E.; Sfravara, F. An experimental comparison between different artificial air cavity designs for a planing hull. *Ocean Eng.* **2017**, *140*, 233–243. [CrossRef]

17.   Cucinotta, F.; Guglielmino, E.; Sfravara, F. A critical CAE analysis of the bottom shape of a multi stepped air cavity planing hull. *Appl. Ocean Res.* **2019**, *82*, 130–142. [CrossRef]

18.   Nguyen, T.; Tran, T.; de Boer, H.; van den Berg, A.; Eilkel, J.C. Rotary-atomizer electric power generator. *Phys. Rev. Appl.* **2015**, *3*, 034005. [CrossRef]

19.   Kirkup, S. The Boundary Element Method in Acoustics: A Survey. *Appl. Sci.* **2019**, *9*, 1642. [CrossRef]

20.   Li, J.D. *Seakeepness of Ships*; Harbin Engineering University: Harbin, China, 1992; ISBN 7-81007-172-6/U.24.

Journal of
Marine Science and Engineering

Article

Evaluation of the Effect of Container Ship Characteristics on Added Resistance in Waves

Ivana Martić [1], Nastia Degiuli [1,*], Andrea Farkas [1] and Ivan Gospić [2]

[1] Faculty of Mechanical Engineering and Naval Architecture, University of Zagreb,Ivana Lučića 5, 10000 Zagreb, Croatia; ivana.martic@fsb.hr (I.M.); andrea.farkas@fsb.hr (A.F.)

[2] The Maritime Department, University of Zadar, Ulica Mihovila Pavlinovića 1, 23000 Zadar, Croatia; igospic@unizd.hr

* Correspondence: nastia.degiuli@fsb.hr

Received: 7 August 2020; Accepted: 6 September 2020; Published: 9 September 2020

Abstract: Added resistance in waves is one of the main causes of an increase in required power when a ship operates in actual service conditions. The assessment of added resistance in waves is important from both an economic and environmental point of view, owing to increasingly stringent rules set by the International Maritime Organization (IMO) with the aim to reduce CO_2 emission by ships. For that reason, it is desirable to evaluate the added resistance in waves already in the preliminary ship design stage both in regular and irregular waves. Ships are traditionally designed and optimized with respect to calm water conditions. Within this research, the effect of prismatic coefficient, longitudinal position of the centre of buoyancy, trim, pitch radius of gyration, and ship speed on added resistance is investigated for the KCS (Kriso Container Ship) container ship in regular head waves and for different sea states. The calculations are performed using the 3D panel method based on Kelvin type Green function. The results for short waves are corrected to adequately take into account the diffraction component. The obtained results provide an insight into the effect of variation of ship characteristics on added resistance in waves.

Keywords: container ship; added resistance in waves; sea states; potential flow theory; variation of ship characteristics

---

1. Introduction

The ship hull is traditionally designed and optimized for calm water conditions, while an increase in the ship resistance owing to sailing in waves is being taken into account via sea margin. Added resistance in waves, as one of the additional loads acting on the ship in service, affects the attainable ship speed, and causes a change in the performance of the ship propulsion system. Its assessment is of great importance as early as in the preliminary design stage to adequately design and optimize the ship propulsion system. The assessment of added resistance ensures the ability of the ship to sail in severe sea states, as well as the possibility to estimate fuel consumption, with the aim of reducing operating costs and emissions of harmful gases. This is of particular importance from the environmental protection point of view and in accordance with regulations introduced by the International Maritime Organization (IMO). IMO has set a series of measures under the EEDI (Energy Efficiency Design Index) and EEOI (Energy Efficiency Operational Indicator) for new built ships and those already in service [1]. When sailing in waves, the total resistance can increase by 15–30% compared with calm water resistance, and still that increase could be even more pronounced [2]. For that reason, it is of particular importance to analyze the effect of variation of ship characteristics on the added resistance in waves. In that way, it is possible to evaluate a change in added resistance for different hull form characteristics rather than optimizing the hull for calm water conditions only. Moreover, it would enable the estimation of the effect of particular loading conditions on added resistance as early as in the preliminary design stage.

The additional power that a ship requires to attain speed when sailing in waves is related to three main components of added resistance. The first and largest one arises owing to the interference of waves generated by the ship response in waves (radiation waves) and incoming waves. It is often called the drift force, although the drift force in the longitudinal direction of the ship is equal to the added resistance in waves only in the case of zero forward speed. The largest contributors to radiation waves are heave and pitch motions and, to a certain extent, roll motion, depending on the phase shift regarding the incoming waves. In short waves, the component of added resistance caused by the ship motions tends to zero, but the second component, called the diffraction force, increases. The wave diffraction is a highly nonlinear phenomenon, thus the accuracy of the methods based on the potential flow theory decreases. The third component of the added resistance is related to the viscous damping of heave and pitch and is negligible compared with the hydrodynamic damping (radiation waves). Therefore, added resistance is considered as an inviscid phenomenon, which allows the application of methods and solvers based on the potential flow theory, as well as an extrapolation of the results from model to full scale, neglecting the scale effects. Added resistance is pressure driven force and can be extrapolated using Froude similarity law. The frictional part of added resistance, subjected to scale effects, is more emphasized in short incoming waves, when it can account for up to 20% of the added resistance. However, it should be emphasized that, in the latter case, added resistance accounts for only up to few percent of calm water resistance [3].

International maritime transport caused an average of 2.6% of global CO_2 emission during the period from 2007 to 2015 [4,5]. For example, in 2015, ships burned about 298 million tons of fuel, of which 72% was heavy fuel oil (HFO), which caused the emission of about 932 million tons of CO_2 into the atmosphere. Assuming that fossil fuels will remain dominant in the future, CO_2 emissions from maritime transport are projected to increase by 50% to 250% by 2050, depending on economic and energy developments [4]. The containerization, which records the most significant increase in cargo transport, is the most energy efficient way to transport cargo and the share of CO_2 emitted by ships is relatively low in global CO_2 emission. However, there is a tendency to improve the energy efficiency of ships and limit greenhouse gas (GHG) emissions as maritime transport continues to evolve and increase. For this reason, IMO has set a goal in 2011 to reduce GHG emissions of ships by at least 50% by 2050 compared with 2008 by introducing mandatory technical measures for new built ships and operational measures for existing ships, with the aim to increase the energy efficiency of ships and reduce CO_2 emissions [1]. EEDI of new built ships is the most important technical measure, aimed at promoting the use of energy efficient equipment and engines, as well as optimization of the propulsion system and ship hull [6]. In the optimization process, it is necessary to take into account the ship response as well as loads acting on the hull in waves. EEDI requires a minimum energy efficiency level per capacity mile for different ship types and sizes. It is expected to encourage innovation and technical development of all components of the ship system that affect fuel consumption, already in the preliminary design phase.

Considering the added resistance in waves as an inviscid phenomenon, numerical methods based on the potential flow theory can be successfully applied for its calculation. Bunnik et al. [7] have analyzed and compared the numerical results of motions and added resistance of the container ship and ferry with the available experimental data. Thus, the authors applied a linear potential flow theory based on the Green function with and without forward speed taken into account, that is, with approximate and exact forward speed, a nonlinear method based on Rankine singularities and computational fluid dynamics (CFD) based on the viscous flow theory. The authors concluded that the application of CFD requires incomparably greater computational resources and does not contribute significantly to the accuracy of the results when nonlinear effects are not pronounced. They also concluded that the methods based on a very similar or identical mathematical model give different results depending on the method of implementation, discretization, or boundary conditions, and that it is more demanding to obtained accurate results for heave than for pitch. Hong et al. [8] have divided the added resistance of S175 container ship into a radiation and diffraction part and applied the Green function based on pulsating and translating sources to solve the boundary condition problem. The applied method provided greater accuracy compared with the experimental results at higher Froude numbers than the Rankine panel method. Linear potential flow theory has been applied to calculate added resistance of intact and damaged S175 ships in [9]. For considered sea states, two calculation models, that is, flooding simulated as increased displacement mass and sloshing inside flooded tank, gave very similar results. Riesner and el Moctar [10] have developed a partially nonlinear time domain method for predicting added resistance at a constant forward speed in regular waves. For the calculation of radiation forces in the time domain, the developed method is based on hydrodynamic coefficients obtained in the frequency domain. The pressure is integrated on the instantaneous wetted surface, which enables the determination of nonlinear Froude–Krylov force and restoring forces. Moreover, an additional force is applied to take into account a variation in wetted surface owing to the radiation and diffraction of incident waves. The additional term, added to take into account the viscous component of the added resistance, is in a form of an exponential function depending on the block coefficient, ratio of wavelength to ship length, and constants determined by the least squares method based on the available results from the literature. Yang et al. [11] have improved the Faltinsen asymptotic formula for calculating the added resistance in short waves. The authors took into account the ship draft, the alteration in the flow velocity around the hull, and the shape of the hull above the waterline. Furthermore, the authors introduced a correction factor associated with a variation in the hull cross section below the waterline, and variation in the hull shape above the waterline was taken into account based on the waterline bluntness coefficient. As new built ships are getting larger, the ratio of wavelength to ship length decreases, meaning that it becomes very important to estimate the added resistance in short waves as accurately as possible. Whether based on potential or viscous flow theory, numerical tools for calculating added resistance in short waves require a large number of panels or finite volumes in order to capture alterations in the flow around the fore part of a hull. Liu and Papanikolaou [12] have developed a relatively simple method for the determination of added resistance in accordance with IMO-MEPC.232(65) EEDI recommendations [13] for establishing the so-called level 1 methods for estimating the minimum required power for navigation and maneuverability of ships in severe weather conditions. The developed empirical method requires only main ship and wave characteristics as input values. The proposed method was extended and adapted for a wider speed range and variation of draft and trim at different loading conditions [14]. The authors analyzed the influence of the advancing speed, the pitch radius of gyration, the draft and the trim on the amplitude and position of the peak value of the added resistance in head waves. They concluded that the peak position is closely related to the natural frequency of heave in long waves and that the amplitude depends on the radius of gyration. Seo et al. [15] have numerically calculated added resistance in short waves using methods based on the potential and viscous flow theory for different ship forms. Given that the added resistance in short waves is greatly influenced by the size of the panel or finite volume, especially for full hull forms, the authors concluded that a convergency study is necessary. Liu and Papanikolaou [16] have utilized the 3D panel method to calculate second-order sway force, that is, drift force for various wave headings at low forward speeds, and concluded that the mean sway force is most significant in relatively short waves. Guha and Falzarano [17] have developed a numerical method for the determination of added resistance based on the direct pressure integration along the instantaneous wetted surface, taking into account the angle of bow flare above the linearized free surface. The authors concluded that it is of great importance to take into account the hull immersion angle when calculating the relative wave amplitude along the waterline on the example of container ships S175 and KCS (Kriso Container Ship), while it does not significantly affect the results of the analyzed Ro-Ro ship. Yang and Kim [18] have come to the similar conclusion by analyzing the bow shape of KVLCC2 (Korean Very Large Crude Carrier) above the waterline. Park et al. [19] have investigated the effect of draft on added resistance in waves experimentally and numerically on the example of KVLCC2. On the basis of the experimental results, the authors concluded that, for short wave lengths, the largest added resistance occurs in the ballast condition and that the peak value shifts towards higher frequencies when reducing the draft. The influence of bow flare on added resistance in waves has been investigated by Fang et al. in [20]. The authors performed numerical calculations, based on the potential flow theory in time domain, and concluded that, by increasing the bow flare angle, the ship motion amplitudes decrease, while added resistance in waves increases. Trim optimization has been performed by various authors for calm water conditions [21,22]. However, the literature lacks studies regarding the added resistance in waves for different trim conditions and trim optimization for sailing in waves. Jung and Kim [23] have applied the multi-objective optimization method to minimize the ship resistance and speed loss caused by wind and waves by varying main dimensions and prismatic coefficient of full KVLCC2 hull form. Kim et al. [24] have proposed a reliable methodology for the estimation of speed loss of the S175 container ship for various sea states and wind conditions. The validation study of ship motions and added resistance obtained using 2D and 3D linear potential flow methods and CFD showed reasonably good agreement with the experimental data in regular head and oblique waves. The authors concluded that the estimated sea margin is significantly increased when the initial reference speed is decreased, mainly owing to added resistance in waves, despite the fact that the absolute value of the required increase in power is lower.

Within this paper, the effect of hull form characteristics of KCS, that is, prismatic coefficient and longitudinal position of the centre of buoyancy, as well as pitch radius of gyration, trim, and sailing speed on added resistance in regular waves and for different sea states is evaluated. To the best of the authors' knowledge, this problem has not been investigated in such a comprehensive way. Considering that added resistance in waves could cause a significant increase compared with calm water resistance, the obtained results provide a valuable insight into the possibility to reduce the added resistance in waves for actual sea states that the ship will encounter during her service. During the design process, certain ship characteristic could be optimal with respect to the calm water resistance, but could disrupt seakeeping characteristics and cause an increase in motion amplitudes or added resistance in waves. The obtained results show the effect of variation of certain ship characteristics on added resistance in waves, on the example of a typical container ship. Added resistance is calculated utilizing the 3D panel method based on the Kelvin type Green function. The numerically obtained results are validated against the experimental data available in the literature and the convergence study is performed to assess the numerical uncertainty.

2. Methodology

2.1. Case Study

Calculations of added resistance in regular and irregular waves are performed for the benchmark container ship, KCS, Figure 1. The main particulars of KCS are shown in Table 1. The original hull form is modified in terms of variation of the prismatic coefficient and the position of the longitudinal centre of buoyancy (LCB) by shifting the hull cross sections in the longitudinal direction, while the midship coefficient and main dimensions remain constant. Cross sections of original and modified hull forms have the same shape and area, but different longitudinal positions. The hull modification method can be applied to ships with and without a parallel mid-body, with the aim of adding a parallel mid-body (if the ship does not have one) or changing its length, while keeping the original value of the prismatic coefficient constant or varying its values [25]. The limits for the variation of the prismatic coefficient and position of LCB are defined according to [25–27]. In order to change the prismatic coefficient of the KCS hull form, cross sections are shifted in the longitudinal direction, without affecting the position of LCB. On the other hand, while changing the position of LCB, an original value of the prismatic coefficient is kept constant.

Figure 1. 3D model of Kriso Container Ship (KCS).

Table 1. The main particulars of Kriso Container Ship (KCS).

Main Particular	
Length between perpendiculars, L_{PP}, m	230
Waterline breadth, B_{WL}, m	32.2
Draught, T, m	10.8
Displacement volume, ∇, m^3	52,030
Block coefficient, C_B	0.651
Midship section coefficient, C_M	0.9849
Longitudinal prismatic coefficient, C_P	0.661
Wetted surface, S, m^2	9530
Longitudinal centre of buoyancy, LCB, %	−1.48
Longitudinal centre of gravity, LCG, m	111.6
Vertical centre of gravity, VCG, m	7.28
Ratio between roll radius of gyration and breadth, r_{xx}/B	0.40
Ratio between pitch radius of gyration and length, r_{yy}/L	0.25
Ship speed, V, kn	24

Cross sections of an example of a modified hull form of KCS with prismatic coefficient equal to 0.688 are shown in Figure 2. In order to analyze the effect of gyration radii on added resistance in waves, the pitch (and yaw) radius of gyration is set as 24%, 25%, and 26% of the length between perpendiculars. The roll radius of gyration is set as 35% of the ship beam. The variation of trim is performed for constant displacement volume with step equal to 0.2° with positive trim defined as bow up and negative trim as bow down. Added resistance in waves for the original hull form of KCS is calculated for three different speeds.

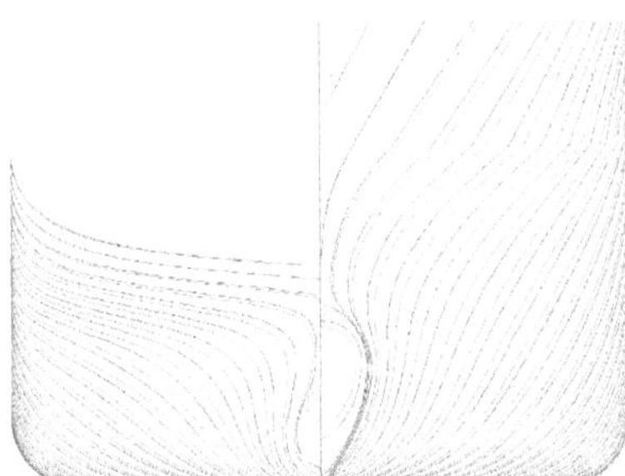

Figure 2. Cross sections of KCS with $C_P = 0.661$ (black line) and $C_P = 0.680$ (blue line).

2.2. Numerical Assessment of Added Resistance in Waves

Within this research, added resistance is numerically calculated utilizing the 3D panel method, by satisfying the Kelvin boundary condition on a free surface. The ship hull is discretized by quadrilateral flat panels with sources of constant and unknown strength located in the collocation points. The velocity field in the computational domain is described by a scalar function of the velocity potential Φ and the Green theorem is applied to obtain the velocity potential by distributing the singularities over the domain boundaries. Boundary integral equations (BIEs) are solved to determine the strength of distributed sources, in a way that the Laplace equation and all the necessary boundary conditions are satisfied. By developing the velocity potential above mean free surface and linearizing the equations under the assumption of small amplitudes and wave steepness, that is, excluding the square terms, the boundary value problem (BVP) of the first order is defined as follows [28]:

$$\nabla \mathbf{V} = \nabla^2 \Phi = 0, \tag{1}$$

$$g\Phi_z + \Phi_{tt} + \mu\Phi_t = 0 \text{ for } z = 0, \tag{2}$$

$$\Phi_n = \mathbf{X}_t\mathbf{n} \quad \text{on wetted surface}, \tag{3}$$

$$\Phi_z = 0 \quad \text{for } z = -h, \tag{4}$$

where $\mathbf{X}_t$ is the velocity vector; $\mathbf{n}$ is the normal vector; and indices t and z denote time and space derivatives, respectively. μ represents a positive and small parameter that introduces the energy dissipation into the momentum equation by a fictitious force dependent on the flow velocity without affecting the inviscid and irrotational properties of the fluid. The result is an additional term in the free surface boundary equation in order to prevent an infinite response at resonant frequencies.

The radiation boundary condition, which requires the velocity potential to disappear at an infinite distance away from the ship hull, is automatically satisfied in the fairly perfect fluid. All velocity potential components satisfy the radiation condition except for the incoming wave potential. Wave elevation ζ evaluated at mean free surface and pressure p on hull are given as follows:

$$\zeta g = -\Phi_t - \mu\Phi, \tag{5}$$

$$\frac{p}{\rho} = -gX_3 - \Phi_t - \mu\Phi, \tag{6}$$

where g is the gravity constant and ρ is the fluid density.

The Green function, $\mathcal{G}$, as a fundamental solution of the Laplace equation, is formulated in such way that it satisfies the linearized boundary condition on the free surface, boundary condition on the bottom, and the radiation condition as follows:

$$\nabla^2\mathcal{G}(P,Q,t) = 4\pi\delta(P-Q), \tag{7}$$

$$g\mathcal{G}_z + \mathcal{G}_{tt} + \mu'\mathcal{G}_t = 0 \quad \text{for } z = 0, \tag{8}$$

$$\mathcal{G}_z = 0 \quad \text{for } z = -h, \tag{9}$$

The Green function represents the field of velocity potential in the computational domain at $P(x,y,z)$ created by the source of unit strength located at $Q(x',y',z')$. $\delta(P-Q) = \delta(x-x')\delta(y-y')\delta(z-z')$ is the Dirac function and $\mu' = \mu$. The free surface Green function is expressed by the Fourier–Hankel integral and approximated by Chebyschev polynomials. More details can be found in [28]. After applying the Green theorem to the domain boundaries (mean free surface, hull surface, sea bed, and cylindrical surface at infinity) and considering a complementary domain inside the ship limited by hull and interior free surface, the integral equation is defined as follows:

$$4\pi\Phi(P) = \iint_H [(\Phi_n - \Phi'_n)\mathcal{G} - (\Phi - \Phi')\mathcal{G}_n]\, ds, \tag{10}$$

where Φ' represents the velocity potential in the interior domain. With the velocity potential and the Green function expressed as $\Phi(P,t) = \text{Re}\{\phi(P)e^{-i\omega t}\}$ and $\mathcal{G}(P,Q,t) = \text{Re}\{G(P,Q)e^{-i\omega t}\}$, respectively; and velocity potential expressed by radiation ϕ_j, diffraction ϕ_7, and incoming wave potential ϕ_0 as $\phi = -i\omega \sum_{j=1}^{6} \zeta_{aj}\phi_j + \zeta_{a0}(\phi_0 + \phi_7)$, with $\phi = \phi'$ and with source strength defined as $\sigma = \phi_n - \phi'_n$.

Moreover, by satisfying the boundary condition on the hull (derivation of the velocity potential in the normal direction), the strength of distributed sources can be determined as follows:

$$2\pi\sigma_j + \iint_H \sigma_j G_n \, ds = \begin{cases} n_j & j = 1 \ldots 6 \\ -\frac{\partial \phi_0}{\partial n} & j = 7 \end{cases}, \tag{11}$$

With the known velocity potential, it is possible to determine the pressure on each panel from the Bernoulli equation. When the point at which the velocity potential is determined is located at a singular point on the hull ($P = Q$), the gradient of the Green function becomes singular. In that case, the small area around the singular point is excluded from the integration. For this reason, a term $2\pi\sigma_j$ is added to Equation (11) that contributes to the amount of normal derivative on the hull in the vicinity of the singular point on the hull. Integral Equation (11) has a unique solution except for the frequencies when the determinant disappears. As the Green function satisfies the free surface boundary condition in the entire computational domain, the solution in the exterior domain of the body is obtained simultaneously as a fictitious solution in the complementary domain. That may cause numerical error at so-called irregular frequencies corresponding to the eigenvalues of the homogenous Dirichlet problem. With $\phi = \phi'$ at these frequencies, the interior BVP has a nontrivial solution. One of the possible solutions to this problem is to define a different boundary condition on the interior free surface, and thus achieve a unique solution for all frequencies. The extended BIEM imposes Neumann's "rigid lid" condition on the interior free surface. In that way, an extended boundary condition is introduced without changing or adding any more parameters to the formulation. For this reason, it is necessary to adequately discretize the interior free surface as well, depending on the frequency range in which the occurrence of the first irregular frequency is expected. The integral equations on the hull and interior free surface read as follows:

$$2\pi\sigma(P) + \iint_{H \cup F'} \sigma(Q)G_n(P,Q)ds = \begin{cases} n_j & j = 1 \ldots 6 \\ -\partial \phi_0/\partial n & j = 7 \end{cases}, \tag{12}$$

$$4\pi\sigma(P) - \iint_{H \cup F'} \sigma(Q)G_n(P,Q)ds = 0, \tag{13}$$

More details on the extended BIEM can be found in [29], where it was shown that the second-order loads are much more sensitive to the irregular frequencies compared with the first-order quantities and that the effects of irregular frequencies should be removed. The second-order wave loads can be determined by integrating the second-order pressure over the mean wetted surface, taking into account the change in the first-order quantities owing to ship motions. In that way, second-order loads can be expressed through one part depending on the variables of the first order and the other part depending on the second-order velocity potential. The ship response to the incoming waves causes a shift from the equilibrium position and a change in the pressure distribution over the wetted surface. Nonlinear forces of a higher order are caused by the change in wetted surface due to the incoming waves, but also as a result of the ship response in waves (especially angular motion), as well as the nonlinear pressure term in the Bernoulli equation. If the force up to the second order is considered, it contains both a constant and an oscillatory part dependent on the square of the wave amplitude. The result of the direct pressure integration along the mean wetted surface and waterline is the quadratic transfer function (QTF) of high-frequency and low-frequency wave loads. The constant drift force, that is, the added resistance, is a time-averaged value of the second-order force, which amounts to about 5% of the first-order wave force and requires only the calculation of the first-order velocity potential. It is represented by the diagonal members of QTF calculated as follows [30,31]:

$$\mathbf{F}_1 = -\frac{1}{2}\rho g \oint_{WL} \zeta_r^2 \mathbf{n} \, dl + \frac{1}{2}\rho \iint_H \nabla\Phi\nabla\Phi \, \mathbf{n} \, ds + \rho \iint_H \mathbf{X}\nabla\Phi_t \mathbf{n} \, ds + mR\ddot{\mathbf{X}}_G, \tag{14}$$

where $\zeta_r = \zeta - X_3$ is the relative wave elevation defined as coupled wave elevation and ship vertical motion and $\mathbf{X} = \mathbf{X}_G + R\mathbf{x}$ is the motion vector of the first order in the ship coordinate system, where the position vector of the point on the hull is given as $\mathbf{x}$ and linearized rotation transformation matrix as R. Equation (14) includes the ship response, the first-order velocity potential, the gradient of the velocity potential, and the wave elevation on mean wetted surface and along the waterline. The first term on the right side of the equation refers to the relative wave elevation, that is, to the integration of the first-order pressure along the variable part of the wetted surface. If the hull emergence angle γ has a non-zero value in the waterline area, that is, the hull surface is not vertical, which is the case for most modern hull forms in the bow and stern area, the normal vector in the integral equation along the waterline needs to be modified according to $\mathbf{n}' = \mathbf{n}/|\cos\gamma|$ [32]. The second and third term in Equation (14) refer to the integration of the first-order pressure along the mean wetted surface, taking into account pressure correction due to ship displacement. The last term refers to the rotation of the first-order force vector and its horizontal component owing to pitch. Namely, first-order hydrodynamic and hydrostatic forces acting along the vertical axis of ship coordinate system will give a longitudinal second-order force component equal to $x_5 F_3 = x_5 m \ddot{X}_{3G}$.

A commercial software package HydroSTAR [33] based on linear potential flow theory is used to perform hydrodynamic calculations. HydroSTAR provides a solution of the first-order wave diffraction and radiation problem, and second-order wave loads with and without forward speed. The immersed part of the ship hull and interior free surface are discretized by quadrilateral panels. Every panel model has approximately 4000 panels on the hull and 6000 panels on the interior free surface. The size of the panels on the interior free surface is equal to 50% of the panel size on the hull in order to shift irregular frequencies further towards a higher frequency range.

The reference coordinate system is located on the mean free surface at the stern with the positive direction of z axis upwards. Within HydroSTAR, mesh is generated using the so-called automatic mesh generator (AMG) based on an adaptive cosine rule, which reduces the panel size from the bottom of the ship towards the waterline, allowing for finer discretization in the area of the waterline. The pressure on the panels closest to the waterline is used to determine the wave elevation using the boundary condition on the free surface. The coordinate system for solving the diffraction and radiation problems is located at the ship centre of buoyancy, and the solution is transformed to the centre of gravity when solving the motion equation.

2.2.1. Correction of the Results in Short Waves

Generally, added resistance in waves calculated by linear potential flow theory significantly underestimates the experimental results for higher incoming wave frequencies. In short waves, where the diffraction force is dominant, linear potential flow theory fails to properly include the diffraction component owing to highly non-linear effects. The results in short waves are thus corrected according to [34] based on the experimental data. With the correction term included, added resistance is calculated as follows:

$$R_{AW} = R_{AWm} + R_{AWr}, \tag{15}$$

where the correction in short waves depends on the bluntness coefficient B_f, draft and incoming wave frequency α_d, and forward speed $(1 + \alpha_U)$ as follows:

$$R_{AWr} = \frac{1}{2}\rho g \zeta_a^2 B B_f \alpha_d (1 + \alpha_U), \tag{16}$$

$$B_f = \frac{1}{B}\left\{ \int_I \sin^2(\alpha + \beta_w)\sin\beta_w dl + \int_{II} \sin^2(\alpha - \beta_w)\sin\beta_w dl \right\}, \tag{17}$$

$$\alpha_d = \frac{\pi^2 I_1^2(k_e T)}{\pi^2 I_1^2(k_e T) + K_1^2(k_e T)}, \tag{18}$$

$$k_e = k\left(1 + \frac{\omega V}{g}\cos\alpha\right)^2, \tag{19}$$

$$1 + \alpha_U = 1 + C_U Fn, \tag{20}$$

$$C_U(\alpha) = \max\left[10, -310B_f(\alpha) + 68\right], \tag{21}$$

where α is the wave heading, β_w is the slope of line element along waterline, I and II are the domains of the integration, I_1 is the first-order Bessel function, K_1 is the second-order Bessel function, k is the wave number, k_e is the encounter wave number, ω is the wave frequency, and C_U is the forward speed coefficient.

2.2.2. Added Resistance in Irregular Waves

The mean value of added resistance in irregular waves can be determined based on the spectral moment m_{0R} or integral of the response spectrum curve. The response spectrum is determined as the product of the added resistance transfer function $\frac{R_{AW}}{\zeta_a^2}(\omega_e)$ in regular waves and the encounter wave energy spectrum $S_\zeta(\omega_e)$ as follows:

$$\overline{R_{AW}} = 2\int_0^\infty S_\zeta(\omega_e)\frac{R_{AW}}{\zeta_a^2}(\omega_e)d\omega_e = 2m_{0R}, \tag{22}$$

where ζ_a is the wave amplitude.

In order to cover the wave energy distributed in the range of encounter frequencies, the wave energy spectrum is transformed into the wave encounter spectrum based on the encounter frequency $\omega_e = \omega - \frac{\omega^2}{g}V\cos\alpha$ as follows:

$$S_\zeta(\omega_e) = \frac{S_\zeta(\omega)}{\sqrt{1 - \frac{4\omega_e V\cos\alpha}{g}}}, \tag{23}$$

It should be noted that Equation (23) is valid only for head and oblique waves. Namely, the following wave encounter spectrum can be double valued because the same encounter frequency corresponds to different incoming wave frequencies. In addition, as the ship speed increases, the wave encounter spectrum extends into the complex solutions range. Within this research, for fully developed sea, a modified two-parameter Pierson–Moskowitz or Bretschneider wave energy spectrum recommended by ITTC (International Towing Tank Conference) is used to calculate added resistance in head waves at different sea states [35]. The Bretschneider wave energy spectrum is defined as follows:

$$S_{\zeta B}(\omega) = \frac{173H_{S1/3}^2/\overline{T}^4}{\omega^5}e^{-\frac{692/\overline{T}^4}{\omega^4}}, \tag{24}$$

With the characteristic wave period expressed through zero crossing period as $\overline{T} = 1,086\,\overline{T}_z$. $H_{S1/3}$ is the significant wave height.

For limited fetch, the modified JONSWAP (Joint North Sea Wave Project) wave energy spectrum is used:

$$S_{\zeta J}(\omega) = \frac{320H_{S1/3}^2}{T_p^4\omega^5}e^{\frac{-1950}{T_p^4\omega^4}}\gamma^{e^{\{-(\frac{\omega/\omega_p-1}{\sigma\sqrt{2}})^2\}}}, \quad \sigma = \begin{cases} 0,07 \text{ za } \omega < \omega_p \\ 0,09 \text{ za } \omega > \omega_p \end{cases}, \tag{25}$$

where $\omega_p = \frac{2\pi}{T_p}$, $\gamma = 3,3$, and peak wave period is defined as $1,073\overline{T}_z = 0,834T_p$.

2.3. Validation and Verification Study

Numerically obtained results of the added resistance in waves are validated against the available experimental data for KCS [36–38] for $Fn = 0.26$, which corresponds to a speed of 24 knots. Relative deviation of the numerical results is calculated as follows:

$$RD = \frac{C_{AW,\text{HSTAR}} - C_{AW,\text{EXP}}}{C_{AW,\text{EXP}}} 100\%, \tag{26}$$

where added resistance coefficient is defined as follows:

$$C_{AW} = \frac{R_{AW}}{\rho g \zeta_a^2 B^2 / L}. \tag{27}$$

The mesh convergence study, as part of the verification procedure [39], is performed to evaluate the effect of the discretization on the obtained numerical results of added resistance, heave, and pitch amplitudes, as well as to quantify the numerical uncertainty. Although it seems that, using a larger number of panels, with sources of constant strength located in collocation points, a more accurate solution can be achieved, this may not be a case. However, a larger number of panels allows more detailed description of the hull geometry, which is of particular importance in the bow and stern area, where significant changes in pressure occur, Figure 3. Generated panel models have 1048, 4366, and 17,823 panels in total. In order to preserve the geometrical similarity between the panel models, each panel is divided into four panels when creating a model with finer discretization.

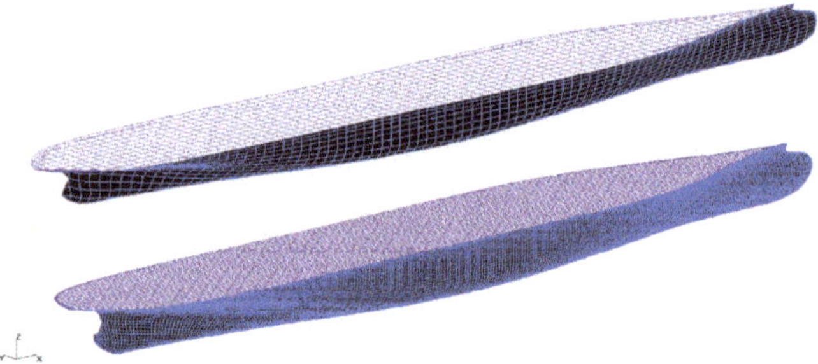

Figure 3. Panel models of KCS: coarse mesh (upper) and fine mesh (lower) [33].

The numerical uncertainty of the obtained results is performed according to the verification procedure recommended by ITTC [40] to estimate the uncertainty of numerical results obtained by viscous flow theory, but it can also be applied to the results of potential flow theory [41]. The numerical uncertainty is evaluated for added resistance coefficient in regular waves and irregular waves by applying the Bretschneider wave energy spectrum in the frequency range from 0.3 to 0.8 rad/s for sea state defined with $H_s = 3.5$ m and $T_Z = 10.5$ s. According to [40], the type of convergence is determined based on the change between solutions obtained using medium and fine mesh $(\phi_2 - \phi_1)$ and coarse and medium mesh $(\phi_3 - \phi_2)$ as follows:

$$R = \frac{(\phi_2 - \phi_1)}{(\phi_3 - \phi_2)}, \tag{28}$$

Monotonic convergence is achieved when $0 < R < 1$, oscillatory convergence when $-1 < R < 0$, and divergence when $|R| > 1$. Numerical uncertainty can be calculated for monotonic convergence case using generalized Richardson extrapolation (RE). The order of accuracy is calculated as follows:

$$p = \frac{\ln((\phi_3 - \phi_2)/(\phi_2 - \phi_1))}{\ln(r)},$$ (29)

where $r = \frac{h_2}{h_1} \cong \frac{h_3}{h_2}$ is the uniform refinement ratio corresponding to the ratio of panel size determined based on the total number of panels as $h_i = 1/\sqrt{N_i}$ [42,43]. With safety factor F_S equal to 1.25, the normalized numerical uncertainty is calculated as follows:

$$\overline{U} = \frac{F_S |\delta_{RE}|}{\phi_1 - \delta_{RE}} \cdot 100,$$ (30)

where RE error is defined as follows:

$$\delta_{RE} = \frac{\phi_2 - \phi_1}{\left(\frac{h_2}{h_1}\right)^p - 1}.$$ (31)

3. Results and Discussion

3.1. Validation of the Numerical Results

The obtained numerical results show satisfactory accuracy compared with the experimental data, except in the short wave region, Figure 4. After the correction of the results for high wave frequencies has been applied, significantly better agreement with the experimental data is obtained. The correction of the results, which accounts for the diffraction part of added resistance, was applied for the results up to $\lambda/L = 0.85$, because it may cause an overestimation of the numerically obtained added resistance values in moderate and long incoming waves. Some discrepancy in the experimental results can be seen as well, especially in the short wave region. By comparing the ship transfer functions of heave and pitch with the experimental data available in the literature, it can be noticed that the pitch transfer function shows much better agreement with the experimental results, with the relative deviation below 5%. The numerically obtained values of heave for moderate and long waves significantly exceed the experimental values. Relative deviation of heave at $\lambda/L = 1474$ is equal to 29%, which confirms that ensuring the accuracy of heave is more demanding than pitch, Figure 5.

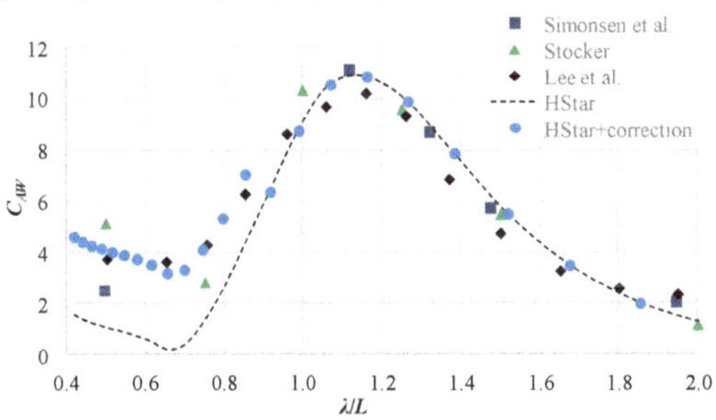

Figure 4. Comparison of the numerically obtained added resistance coefficient with available experimental data [36–38].

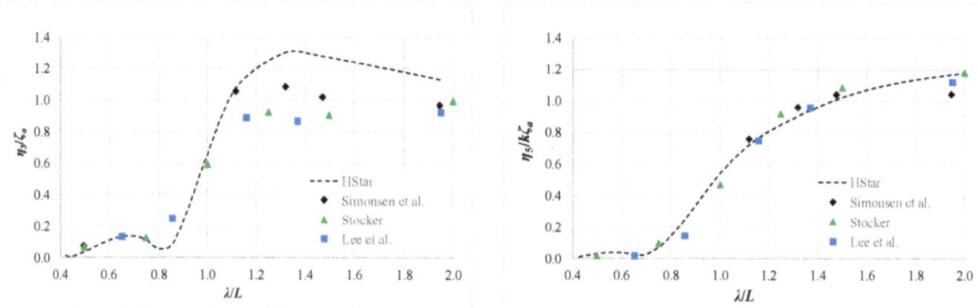

Figure 5. Comparison of the numerically obtained heave (**left**) and pitch (**right**) transfer functions with available experimental data [36–38].

3.2. Verification of the Numerical Results

Table 2 shows the hydrostatic characteristics of KCS obtained using panel models with different mesh density. The differences between the hydrostatic characteristics for different panel models are not significant, even though the displacement volume of the finest discretization is closest to the actual value. It should be noted that the position of the vertical centre of buoyancy is given with the respect to the coordinate system located on the mean free surface. The effect of panel model discretization on the obtained results of heave and pitch motion amplitudes can be seen in Figure 6. Increasing the number of panels does not influence the obtained results, unlike in the case of added resistance. The largest change in the results is obtained for the highest frequency using fine mesh. Without the correction being applied, this value has the lowest relative deviation compared with the experimental results. The numerical uncertainty of the obtained added resistance coefficient in regular waves and coefficient of the mean added resistance in irregular waves can be seen in Table 3. It can be seen that the calculated numerical uncertainty, for the results with monotonic convergence, is below 2.1% for regular and equal to 1.64% for irregular waves.

Table 2. Hydrostatic characteristics of KCS obtained using different panel models.

N_i	1048	4366	17823
∇, m^3	51690	51884	51930
LCB, m	111.555	111.594	111.605
VCB, m	−4.876	−4.886	−4.888
S, m^2	9479.1	9513.2	9520.2
A_{WL}, m^2	6144.8	6141.2	6139.0

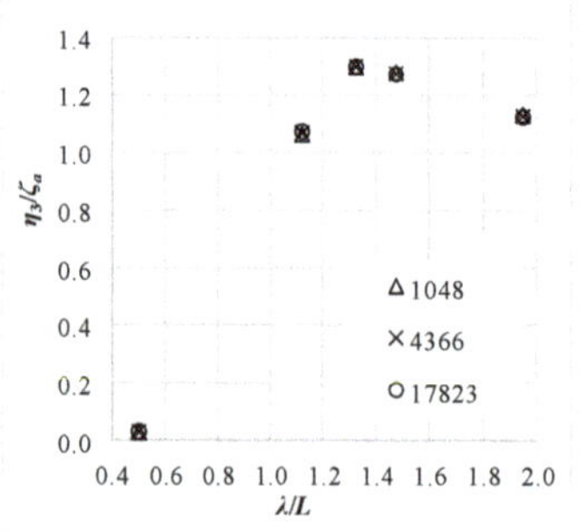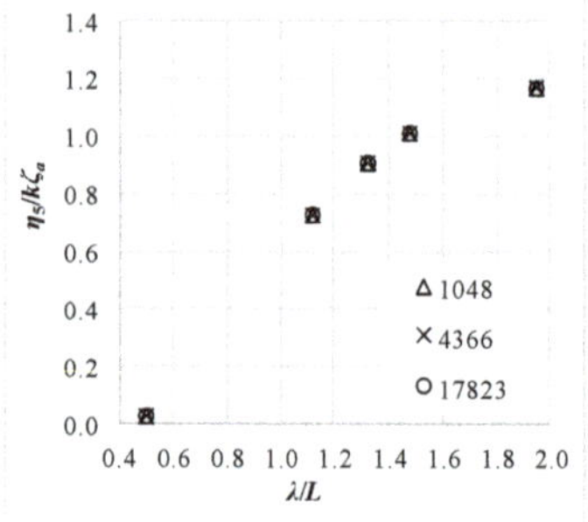

Figure 6. Heave (**left**) and pitch (**right**) transfer functions obtained using different panel models.

Table 3. Convergence study for panel size used to discretize KCS.

N_i	1048	4366	17823					
h_i	0.0309	0.0151	0.0075					
λ/L	ϕ_3	ϕ_2	ϕ_1	R	p	δ_{RE}	ϕ_0	$\overline{U}$, %
2.0	1.518	1.513	1.492	4.4882	/	/	/	/
1.5	6.326	6.244	6.221	0.2911	1.7803	0.0098	6.211	0.197
1.33	9.193	9.069	8.989	0.6483	0.6252	0.1484	8.840	2.098
1.15	11.279	10.996	10.718	0.9828	/	/	/	/
0.5	0.780	0.762	1.568	−43.1464	/	/	/	/
C_{AW}	10.813	10.578	10.455	0.5238	0.9329	0.1353	10.320	1.639

3.3. Results of the Sensitivity Analysis

The effect of the prismatic coefficient of KCS on added resistance in regular and irregular waves (Table 4) is presented by the results in Table 5. The results are given as relative deviations with respect to the results obtained for the original hull form of KCS. It can be seen that, in long and moderate regular waves, except for the wave frequency corresponding to the peak value of added resistance, the increase in prismatic coefficient and block coefficient at the same time causes the decrease in added resistance. Namely, for large wavelengths, without the presence of diffraction component, an increase in the prismatic coefficient leads to smaller amplitudes of the ship motions. The relative motions become large when the length of the ship is approximately equal to the wavelength, causing the largest resistance. In very long waves, the relative motions tend to zero, and the force tends to a Froude–Krylov force that would ideally act on the ship in the absence of a diffraction component. Consequently, in very long waves, the added resistance tends to zero. The peak value of the added resistance decreases with the decrease in the prismatic coefficient. In order to analyze the effect of variation of the prismatic coefficient on added resistance in short waves, a correction for diffraction was applied. It can be noticed that the diffraction part of added resistance in short waves increases as the prismatic coefficient increases, which means that the fore part of ship is fuller. For irregular waves, the mean value of added resistance was calculated for six sea states by means of spectral analysis based on two theoretical wave energy spectra, Table 4. Spectral analysis based on the corrected results showed that added resistance in irregular waves decreases for hull modifications with a lower value of the prismatic coefficient, and vice versa, and that a change in added resistance is more pronounced for lower sea states when wave energy is concentrated in the high frequency range. In that way, the diffraction component of added resistance determines the trend of mean value in irregular waves. At higher sea states, with wave energy distributed along the entire frequency range, the obtained results are mainly within numerical uncertainty and the results in regular waves counteract each other.

Table 4. Significant wave height and wave period for considered sea states.

Sea State	H_S, m	T_Z, s
SS1	1.5	6.5
SS2	2.5	7.5
SS3	2.5	8.5
SS4	3.5	9.5
SS5	3.5	10.5
SS6	4.5	11.5

Table 5. Influence of the prismatic coefficient on added resistance in regular and irregular waves. JONSWAP, Joint North Sea Wave Project.

C_P	0.629	0.64	0.65	0.67	0.68	0.69	0.701
λ/L				*RD*, %			
2.0	4.65	1.64	0.49	−3.85	−5.76	−6.85	−9.73
1.5	4.52	1.62	0.52	−3.11	−5.06	−6.48	−8.33
1.33	3.04	0.58	−0.02	−2.88	−4.55	−5.36	−7.06
1.15	−3.47	−3.30	−1.82	−0.01	−0.32	1.73	2.74
0.5	−6.73	0.26	0.83	4.76	4.68	7.48	8.89
				RD, % (Bretschneider)			
SS1	−5.36	−1.04	−0.16	2.88	2.68	5.15	6.36
SS2	−3.57	−2.18	−1.08	0.76	0.38	2.32	3.19
SS3	−2.64	−2.27	−1.24	−0.03	−0.55	1.07	1.67
SS4	−2.08	−2.14	−1.22	−0.39	−1.02	0.39	0.81
SS5	−1.45	−1.95	−1.18	−0.57	−1.18	−0.27	0.31
SS6	−1.44	−1.88	−1.12	−0.74	−1.49	−0.34	−0.16
				RD, % (JONSWAP)			
SS1	−6.61	0.14	0.73	4.59	4.50	7.27	8.66
SS2	−4.99	−1.28	−0.36	2.44	2.20	4.56	5.70
SS3	−3.28	−2.35	−1.22	0.43	0.01	1.88	2.68
SS4	−2.94	−2.77	−1.53	−0.10	−0.54	1.25	2.02
SS5	−2.10	−2.41	−1.40	−0.44	−0.98	0.34	0.99
SS6	−0.36	−1.28	−0.84	−1.20	−2.17	−1.48	−1.69

In Table 6, the obtained results of the effect of the longitudinal centre of buoyancy on added resistance in regular and irregular waves are shown. In order to change the position of *LCB*, for example, towards the bow, cross sections of the fore part were moved towards bow and cross sections of the aft part towards midship. In that way, the fore prismatic coefficient increases, while the aft one decreases, keeping the original value of the total prismatic coefficient constant. Two *LCB* positions are analyzed and the obtained numerical results are given as relative deviations with regard to results of the original KCS hull form. On the basis of the obtained numerical results, it is possible to conclude that the position of *LCB* does not significantly affect added resistance in long regular waves. It can be seen that the peak value of added resistance decreases by almost 6% with *LCB* shifted towards stern, and increases by over 3% with *LCB* shifted towards bow. In general, shifting the position of *LCB* towards the stern allows a fine shape of the fore part of the ship. On the other hand, shifting the position of *LCB* towards the bow increases the distance between the longitudinal centre of buoyancy and longitudinal centre of floatation, which reduces the ship relative motions, causing the decrease in added resistance in relatively short waves. This can be observed for $\lambda/L = 0.5$, where added resistance increases when *LCB* shifts towards stern. A noticeable effect of *LCB* on the mean value of added resistance in irregular waves can be observed for higher sea states, when the position of *LCB* is shifted towards the stern. The decrease in added resistance in that case equals to 3.5–5% for both theoretical wave energy spectra.

To perform seakeeping calculations, it is necessary to know the mass characteristics along with the hull form characteristics. As shown in [44], the influence of the vertical position of the centre of gravity on added resistance in head waves is almost negligible when calculated using linear potential flow theory and by keeping the values of gyration radii constant. Changing the position of the centre of gravity does not affect the terms of hydrodynamic mass matrix, given that the coordinate system is placed at the centre of gravity. On the other hand, an insignificant change in terms of hydrodynamic added mass and damping matrix, especially in the case of heave and pitch, is noticed. In the restoring forces matrix, the large value of longitudinal metacentric height changes only slightly by shifting the vertical position of the centre of gravity.

Table 6. Influence of the position of the longitudinal centre of buoyancy on added resistance in regular and irregular waves.

LCB, m	111 m	113 m
λ/L	*RD*, %	
2.0	−2.38	0.89
1.5	−0.62	−0.82
1.33	−2.32	0.09
1.15	−5.74	3.28
0.5	3.75	−2.24
	RD, % (Bretschneider)	
SS1	0.72	−0.49
SS2	−3.31	1.77
SS3	−4.15	2.17
SS4	−4.27	2.18
SS5	−4.27	2.14
SS6	−4.19	2.06
	RD, % (JONSWAP)	
SS1	3.52	−2.11
SS2	−0.50	0.20
SS3	−3.87	2.08
SS4	−4.99	2.73
SS5	−4.81	2.53
SS6	−3.65	1.57

Pitch gyration radius in head waves has a more significant effect on added resistance in waves. As mentioned before, its value is varied as a percentage of ship length between perpendiculars. The initial value of pitch (and yaw) gyration radius for KCS container ship corresponds to 25% of the length between perpendiculars. Additionally, values corresponding to 24% and 26% of the length between perpendiculars are analyzed, Figure 7. It can be seen that, by increasing and decreasing thpitch gyration radius, added resistance in long and moderate waves increases and decreases, respectively, by approximately 15%, while the change in peak value amounts to about 6%. According to the obtained numerical results in short waves, there is no effect of change on the pitch gyration radius, which is expected considering that the amplitudes of absolute ship motions in that frequency range are small. By changing the pitch radius of gyration, it can be seen from Figure 8 that added resistance in irregular waves decreases or increases by up to 10% for higher sea states. The change in added resistance is more pronounced for the Bretchneider than for the JONSWAP sea spectrum, except for SS6, when the wave energy described by the Bretschneider spectrum shifts towards lower wave frequencies. The narrow banded JONSWAP spectrum has a higher peak value in the range of high wave frequencies, causing a larger change in mean added resistance for SS1, even though the mean value of added resistance in that case equals approximately 20 kN.

Figure 9 shows the obtained numerical values of added resistance in regular waves for three speeds: 20, 22, and 24 knots, which correspond to Froude numbers of 0.216, 0.238, and 0.260, respectively. Compared with the design speed equal to 24 knots, at the frequency corresponding to the peak value of added resistance, a decrease of speed by 2 knots causes a decrease in added resistance by 15%, and the decrease by 4 knots causes the decrease by an additional 14%. This decrease is even more pronounced at lower wave frequencies. For example, a decrease in added resistance by decreasing the speed by 2 and 4 knots is about 15% and 28% for $\lambda/L = 1.33$ and 22% and 36% for $\lambda/L = 1.5$, respectively. It can also be seen that, as the speed increases, the peak value of the added resistance slightly shifts towards lower frequencies. The corrected results in short waves show a significant decrease in added resistance for lower speeds: 25% and 35% for 22 knots and 20 knots, respectively.

contribute significantly to the decrease in added resistance. The largest change in added resistance calculated by means of both wave energy spectra can be observed for SS1 owing to the diffraction component. It should be noted that some of the obtained numerical results are within the calculated numerical uncertainty.

Table 7. Influence of trim on added resistance in regular and irregular waves.

$t,°$	0.8	0.6	0.4	0.2	−0.2	−0.4	−0.6	−0.8	−1
λ/L					$RD, \%$				
2.0	15.07	−7.41	−3.29	1.94	−2.84	−5.00	−4.08	−3.91	−5.00
1.5	4.92	−7.34	−3.80	−0.09	0.04	−1.41	0.29	1.80	1.89
1.33	4.21	−6.42	−3.12	0.44	0.24	−1.36	0.66	2.51	3.06
1.15	12.05	−0.23	0.03	0.79	0.07	−1.87	−0.72	1.47	3.16
0.5	−18.09	−17.08	−10.46	−8.48	8.59	10.01	12.45	10.03	2.60
				$RD, \%$ (Bretschneider)					
SS1	−4.58	−7.37	−4.46	−3.39	3.66	3.75	5.10	4.64	1.93
SS2	4.23	−3.81	−2.18	−1.08	1.65	0.63	1.92	2.92	2.59
SS3	7.42	−2.72	−1.43	−0.18	0.84	−0.61	0.69	2.24	2.80
SS4	8.37	−2.56	−1.29	0.13	0.54	−1.06	0.27	1.99	2.81
SS5	8.70	−2.61	−1.29	0.26	0.39	−1.26	0.08	1.86	2.76
SS6	8.84	−2.71	−1.32	0.33	0.31	−1.37	−0.02	1.77	2.69
				$RD, \%$ (JONSWAP)					
SS1	−9.95	−9.63	−5.89	−4.76	4.84	5.60	7.00	5.68	1.53
SS2	−3.05	−6.75	−4.06	−2.99	3.31	3.21	4.55	4.34	2.05
SS3	6.02	−3.09	−1.72	−0.61	1.24	0.00	1.28	2.56	2.71
SS4	9.90	−1.45	−0.68	0.37	0.40	−1.33	−0.09	1.81	2.99
SS5	9.66	−1.94	−0.90	0.47	0.27	−1.49	−0.18	1.77	2.97
SS6	8.03	−3.50	−1.73	0.30	0.27	−1.38	0.07	1.86	2.69

The effect of ship characteristics on added resistance in regular head waves depends on the frequency of the incoming wave. By increasing the prismatic coefficient to the highest value considered within this study, added resistance in short waves and the peak value of added resistance increase, while in moderate and long waves, added resistance decreases. In very long waves, the change in the prismatic coefficient as well as the change in the position of the longitudinal centre of buoyancy have a negligible effect. The greatest effect of variation of the LCB position is observed at the peak value of the added resistance. As the speed decreases, the added resistance decreases for all analyzed λ/L values. Finally, as the pitch gyration radius increases, the value of the added resistance increases as well, more significantly at large and moderate wavelengths. Depending on the frequency range covered by spectral energy, for certain sea states, the influence of the pitch gyration radius will be negligible, while for the other sea states with wave energy distributed in the area of lower frequencies, this influence will be significant. A similar trend can be observed in the case of other varied characteristics. It can be concluded that the results are highly dependent on the incoming wave frequencies and characteristics of sea states, as well on the applied theoretical wave energy spectrum. The distribution of wave energy in the frequency range is dependent on the wave period and the intensity of change in added resistance by varying different ship characteristics on the wave height.

4. Conclusions

Hydrodynamic calculations of added resistance of KCS in regular waves were performed in the frequency range utilizing the 3D panel method based on the potential flow theory. The obtained numerical results were corrected for the diffraction component of added resistance in short waves. The numerical results were validated against the experimental data available in the literature and satisfactory agreement was achieved. Numerical uncertainty was evaluated for those results in regular waves, whereas monotonic convergence was achieved and for the mean value of added resistance in irregular waves for certain sea states. At high wave frequencies, there is a potential problem of the appearance of irregular frequencies when calculating second-order forces and the results are unreliable in this area. In that frequency range, the results depend mostly on the applied correction based on the waterline shape, speed, draft, and wave characteristics. Within this research, the effect of ship characteristics on added resistance in regular waves and for certain sea states was investigated. By increasing the prismatic coefficient, while keeping the main dimensions and midship coefficient constant, the added resistance in long and moderate regular waves decreases, except for the peak value, owing to the smaller amplitudes of ship motions. In short waves, the change in added resistance due to the change in the prismatic coefficient is more pronounced. Namely, with the increase in the prismatic coefficient, added resistance, mostly caused by the diffraction of the waves, increases. Spectral analysis based on the corrected results showed that, despite the decrease in the motion amplitudes, the mean value of added resistance in irregular waves for lower sea states increases with the increase in the prismatic coefficient. On the basis of the results obtained by shifting the *LCB* position, the largest effect on added resistance in regular waves was observed for the peak value. Shifting *LCB* towards the bow increases the distance between the longitudinal centre of buoyancy and longitudinal centre of floatation, reducing the amplitudes of relative motions. The mean value of the added resistance in higher sea states decreases by shifting *LCB* towards the stern. As already known, by decreasing the ship speed, a significant reduction in added resistance can be achieved. A significant influence on added resistance was observed for the pitch gyration radius as well, especially for moderate wavelengths. Additionally, sensitivity analysis of trim was performed considering that optimum trim can lead to higher ship efficiency, not only in calm water, but in waves as well. It was observed that trim by stern up to 0.6° reduces added resistance for actual sea states by 2–3%.

As the effect of different ship characteristics depends on the distribution of wave energy in the frequency range, the obtained results provide a valuable insight into the effect of the variation of ship characteristics on added resistance for a particular sea state. When designing or optimizing a ship and its propulsion system for calm water conditions, it is important to evaluate the effect of the variation of ship characteristics on added resistance in waves as well, as it is one the main causes of an increase in required power in service. Ship characteristics optimized for calm water performance are not necessarily optimal for real sailing conditions. Considering that a ship in service encounters waves from various directions, it would be beneficial to investigate the effect of variations in ship characteristics for different headings. This will form a part of future research.

Author Contributions: Conceptualization, I.M. and N.D.; methodology, I.M., N.D., A.F., and I.G.; software, I.M.; validation, I.M.; formal analysis, I.M.; investigation, I.M., N.D., A.F., and I.G.; resources, I.M.; writing—original draft preparation, I.M., N.D., A.F., and I.G.; writing—review and editing, I.M., N.D., A.F., and I.G.; visualization, I.M; supervision, N.D. All authors have read and agreed to the published version of the manuscript.

Funding: This research received no external funding.

Conflicts of Interest: The authors declare no conflict of interest.

References

1. International Maritime Organization. *MARPOL Annex VI MEPC*; IMO: London, UK, 2013; Volume 203.
2. Perez Arribas, F. Some methods to obtain the added resistance of a ship advancing in waves. *Ocean Eng.* **2007**, *34*, 946–955. [CrossRef]
3. Sigmund, S.; el Moctar, O. Numerical and experimental investigation of added resistance of different ship types in short and long waves. *Ocean Eng.* **2018**, *147*, 51–67. [CrossRef]
4. International Maritime Organization. *Third IMO Greenhouse Gas Study 2014*; IMO: London, UK, 2014.
5. Olmer, N.; Comer, B.; Roy, B.; Mao, X.; Rutherford, D. *Greenhouse Gas Emissions from Global Shipping, 2013–2015*; ICCT: Washington DC, USA, 2017.
6. Ren, H.; Ding, Y.; Sui, C. Influence of EEDI (Energy Efficiency Design Index) on Ship–Engine–Propeller Matching. *J. Mar. Sci. Eng.* **2019**, *7*, 425. [CrossRef]

7.   Bunnik, T.; van Daalen, E.; Kapsenberg, G.; Shin, Y.; Huijsmans, R.; Deng, G.; Delhommeau, G.; Kashiwagi, M.; Beck, B. A comparative study on state-of-the-art prediction tools for seakeeping. In Proceedings of the 28th Symposium on Naval Hydrodynamics, Pasadena, CA, USA, 12–17 September 2010.

8.   Hong, L.; Zhu, R.; Miao, G.; Fan, J.; Li, S. An investigation into added resistance of vessels advancing in waves. *Ocean Eng.* **2016**, *123*, 238–248. [CrossRef]

9.   Martić, I.; Degiuli, N.; Ćatipović, I. Added resistance in waves of intact and damaged ship in the Adriatic Sea. *Brodogradnja.* **2015**, *66*, 1–14.

10. Riesner, M.; Moctar, O. A time domain boundary element method for wave added resistance of ships taking into account viscous effects. *Ocean Eng.* **2018**, *162*, 290–303. [CrossRef]

11. Yang, K.; Kim, Y.; Jung, Y. Enhancement of asymptotic formula for added resistance of ships in short waves. *Ocean Eng.* **2018**, *148*, 211–222. [CrossRef]

12. Liu, S.; Papanikolaou, A. Fast approach to the estimation of the added resistance of ships in head waves. *Ocean Eng.* **2016**, *112*, 211–225. [CrossRef]

13. International Maritime Organization. *Interim Guidelines for Determining Minimum Propulsion Power to Maintain the Manoeuvrability of Ships in Adverse Conditions—Resolution MEPC*; IMO: London, UK, 2013; Volume 232.

14. Liu, S.; Papanikolaou, A. Approximation of the added resistance of ships with small draft or in ballast condition by empirical formula. *Proc. Inst. Mech. Eng. Part M: J. Eng. Marit. Environ.* **2017**, *233*, 27–40. [CrossRef]

15. Seo, M.; Yang, K.; Park, D.; Kim, Y. Numerical analysis of added resistance on ships in short waves. *Ocean Eng.* **2014**, *87*, 97–110. [CrossRef]

16. Liu, S.; Papanikolaou, A. Prediction of the Side Drift Force of Full Ships Advancing in Waves at Low Speeds. *J. Mar. Sci. Eng.* **2020**, *8*, 377. [CrossRef]

17. Guha, A.; Falzarano, J. The effect of hull emergence angle on the near field formulation of added resistance. *Ocean Eng.* **2015**, *105*, 10–24. [CrossRef]

18. Yang, K.; Kim, Y. Numerical analysis of added resistance on blunt ships with different bow shapes in short waves. *J. Mar. Sci. Technol.* **2017**, *22*, 245–258. [CrossRef]

19. Park, D.; Kim, Y.; Seo, M.; Lee, J. Study on added resistance of a tanker in head waves at different drafts. *Ocean Eng.* **2016**, *111*, 569–581. [CrossRef]

20. Fang, M.; Lee, Z.; Huang, K. A simple alternative approach to assess the effect of the above-water bow form on the ship added resistance. *Ocean Eng.* **2013**, *57*, 34–48. [CrossRef]

21. Sun, J.; Tu, H.; Chen, Y.; Xie, D.; Zhou, J. A study on trim optimization for a container ship based on effects due to resistance. *J. Ship Res.* **2016**, *60*, 30–47. [CrossRef]

22. Lyu, X.; Tu, H.; Xie, D.; Sun, J. On resistance reduction of a hull by trim optimization. *Brodogradnja* **2018**, *69*, 1–13. [CrossRef]

23. Jung, Y.-W.; Kim, Y. Hull form optimization in the conceptual design stage considering operational efficiency in waves. *Proc. Inst. Mech. Eng. Part M: J. Eng. Marit. Environ.* **2019**, *233*, 745–759. [CrossRef]

24. Kim, M.; Hizir, O.; Turan, O.; Day, S.; Incecik, A. Estimation of added resistance and ship speed loss in a seaway. *Ocean Eng.* **2017**, *141*, 465–476. [CrossRef]

25. Lackenby, H. On the systematic geometrical variation of ship forms. *Trans. Inst. Nav. Archit.* **1950**, *92*, 289–315.

26. Papanikolaou, A. *Ship Design, Methodologies of Preliminary Design*; Springer: Berlin/Heidelberg, Germany, 2014.

27. Kristensen, H.O. *Statistical Analysis and Determination of Regression Formulas for Main Dimensions of Container Ships based on IHS Fairplay Data*; University of Southern Denmark: Odense, Denmark, 2013.

28. Chen, X. Hydrodynamics in Offshore and Naval Applications—Part I. An updated version of the paper presented as a keynote lecture. In Proceedings of the 6th International Conference on Hydrodynamics, Perth, Australia, 24–26 November 2004; pp. 1–28.

29. Martić, I.; Degiuli, N.; Malenica, Š.; Farkas, A. Discussions on the Convergence of the Seakeeping Simulations Based on the Panel Methods. In Proceedings of the International Conference on Offshore Mechanics and Arctic Engineering, OMAE 2018, Madrid, Spain, 17–22 June 2018.

30. Pinkster, J.A. Mean and low frequency wave drifting forces on floating structures. *Ocean Eng.* **1979**, *6*, 593–615. [CrossRef]

31. Journée, J.M.J.; Massie, W.W. *Offshore Hydromechanics*; TU Delft: Mekelweg, The Netherlands, 2001.

32. Chen, X.B. Middle-field formulation for the computation of wave drift loads. *J. Eng. Math.* **2007**, *59*, 61–82. [CrossRef]

33. *Hydrostar for Experts User Manual*; Bureau Veritas: Paris, France, 2016.

34. Tsujimoto, M.; Kuroda, M.; Shibata, K.; Takagi, K. A Practical Correction Method for Added Resistance in Waves. *J. Jpn. Soc. Nav. Archit. Ocean Eng.* **2008**, *8*, 177–184. [CrossRef]

35. ITTC. Final report and recommendations to the 22nd ITTC. In Proceedings of the 22nd International Towing Tank Conference (ITTC'99), Seoul, Korea & Shanghai, China, 5–11 September 1999.

36. Simonsen, C.D.; Otzen, J.F.; Joncquez, S.; Stern, F. EFD and CFD for KCS heaving and pitching in regular head waves. *J. Mar. Sci. Technol.* **2013**, *18*, 435–459. [CrossRef]

37. Lee, C.; Park, S.; Yu, J.; Choi, J.; Lee, I. Effects of diffraction in regular head waves on added resistance and wake using CFD. *Int. J. Nav. Arch. Ocean Eng.* **2019**, *11*, 736–749. [CrossRef]

38. Stocker, M.R. *Surge Free Added Resistance Tests in Oblique Wave Headings for the KRISO Container Ship Model*; University of Iowa: Iowa City, IA, USA, 2016.

39. ITTC. *ITTC—Recommended Procedures and Guidelines, Verification and Validation of Linear and Weakly Nonlinear Seakeeping Computer Codes 7.5–02-07-02.5*; ITTC: Zürich, Switzerland, 2011.

40. ITTC. *ITTC—Recommended Procedures and Guidelines, Uncertainty Analysis in CFD Verification and Validation, Methodology and Procedures 7.5-03 -01-01*; ITTC: Zürich, Switzerland, 2017.

41. Hizir, O.; Kim, M.; Turan, O.; Day, A.; Incecik, A.; Lee, Y. Numerical studies on non-linearity of added resistance and ship motions of KVLCC2 in short and long waves. *Int. J. Nav. Arch. Ocean Eng.* **2019**, *11*, 143–153. [CrossRef]

42. Eca, L.; Hoekstra, M. An Evaluation of Verification Procedures for CFD Applications. In Proceedings of the 24th Symposium on Naval Hydrodynamics, Fukuoka, Japan, 8–13 July 2002.

43. Eca, L.; Vaz, G.B.; de Campos, J.A.C.F.; Hoekstra, M. A verification of calculations of the potential flow around 2D foils. *AIAA J. Am. Inst. Aeronaut. Astronaut.* **2004**, *42*, 2401–2407. [CrossRef]

44. Martić, I.; Degiuli, N.; Komazec, P.; Farkas, A. Influence of the approximated mass characteristics of a ship on the added resistance in waves. In Proceedings of the 17th International Congress on Maritime Transportation and Harvesting of Sea Resources, IMAM 2017, Lisbon, Portugal, 9–11 October 2017; pp. 439–446.

Journal of
Marine Science and Engineering

Article

On the Comparative Seakeeping Analysis of the Full Scale KCS by Several Hydrodynamic Approaches

Florin Pacuraru, Leonard Domnisoru and Sandita Pacuraru *

Department of Naval Architecture, "Dunarea de Jos" University of Galati, 800008 Galati, Romania;
florin.pacuraru@ugal.ro (F.P.); leonard.domnisoru@ugal.ro (L.D.)
* Correspondence: sorina.pacuraru@ugal.ro

Received: 17 October 2020; Accepted: 19 November 2020; Published: 25 November 2020

Abstract: The main transport channel of the global economy is represented by shipping. Engineers and hull designers are more preoccupied in ensuring fleet safety, the proper operation of the ships, and, more recently, compliance with International Maritime Organization (IMO) regulatory incentives. Considerable efforts have been devoted to in-depth understanding of the hydrodynamics mechanism and prediction of ship behavior in waves. Prediction of seakeeping performances with a certain degree of accuracy is a demanding task for naval architects and researchers. In this paper, a fully numerical approach of the seakeeping performance of a KRISO (Korea Research Institute of Ships and Ocean Engineering, Daejeon, South Korea) container ship (KCS) container vessel is presented. Several hydrodynamic methods have been employed in order to obtain accurate results of ship hydrodynamic response in regular waves. First, an in-house code DYN (Dynamic Ship Analysis, "Dunarea de Jos" University of Galati, Romania), based on linear strip theory (ST) was used. Then, a 3D fully nonlinear time-domain Boundary Element Method (BEM) was implemented, using the commercial code SHIPFLOW (FLOWTECH International AB, Gothenburg, Sweden). Finally, the commercial software NUMECA (NUMECA International, Brussels, Belgium) was used in order to solve the incompressible unsteady Reynolds-averaged Navier–Stokes equation (RANSE) flow at ship motions in head waves. The results obtained using these methods are represented and discussed, in order to establish a methodology for estimating the ship response in regular waves with accurate results and the sensitivity of hydrodynamical models.

Keywords: strip theory; BEM; RANS; regular wave; seakeeping

---

1. Introduction

Shipping, often considered as the main transport channel of the global economy, is responsible for approximately 80 percent of world trade. An increasing focus on more environmentally friendly shipping pushes hull designers to further ensure fleet safety, proper operation of the ships, and, more recently, compliance with International Maritime Organization (IMO) regulatory incentives regarding the energy efficiency operational indicator (EEOI) and emissions reduction. Considerable efforts have been devoted to an in-depth understanding of the hydrodynamics mechanism and the prediction of ship behavior in waves. In addition to safety, efficiency and operability, the waves' loads may determine structural failure. Prediction of seakeeping performances with a certain degree of accuracy is a demanding task for naval architects and of great practical interest for shipbuilders, owners, operators, as it affects both the ships' design and operation [1]. More recently, the shipping industry has embraced more and more digitalization. Digital twin concept couples physical–numerical modelling and simulation to evolve solutions that improve vessel predictability, behavior control and response.

There are several aspects concerning seakeeping that make it one of the most challenging problems in ship hydrodynamics. The nonlinearities induced by the fluid viscosity and hydrodynamic pressure, boundary conditions at free-surface formulation, interference between external waves and ship's body, specific geometry of the hull shape, all require advanced iterative time-domain procedures for the ship's oscillations response analysis, involving significant computational resources [2].

The problem of a moving body interacting with waves has been pursued since the very first Froude and Michel studies. Since then, different approaches, such as experimental fluid dynamics, potential flow and recently computational fluid dynamics, have been developed to estimate the seakeeping performances. Traditionally, the hydrodynamic performances of a full-scale ship are determined by extrapolating the results of model-scale towing tank tests. Another approach commonly adopted in ship hydrodynamics is the employment of numerical techniques to solve the equations' system of ship motions. The prediction of non-linear phenomena specific to ship motions in waves is difficult to be solved numerically, not only due to the complexity of the problem, but also the results are highly dependent on the details of the hull form and the incident wave condition. Due to the development of numerical modelling methods and computing power increase, direct prediction of full-scale ship performance became a practical approach. Most of the available techniques used to predict ship motions rely on assumptions from potential flow theory. Even if the natural trend in ship hydrodynamics is to move from frequency-domain to time-domain approach, from linear 2D strip theory type to fully 3D nonlinear techniques, and from potential to viscous computation, potential flow methods are still highly used in ship design, since they provide robust and quite accurate results in low to moderate sea states [3].

The frequency-domain approach is carried out mainly for linear or weakly nonlinear wave theories, where time dependence can be removed by assuming that solution is harmonic in time, in consequence only steady solutions are solved. Assuming that the ship body is slender, strip theory simplifies the 3D flow problem into a 2D formulation, modelling the ship hull as a set of multiple 2D ship stations [4]. The independent boundary value problem for each station can be solved analytically (Lewis form method) or by boundary element method. For the 2D hydrodynamic formulation, since the 1950s, various strip theories have been developed for the seakeeping problem [5]. The pioneer work of Korvin-Kroukovsky [6] set the principal feature of the strip theory for calculating ship motions based on the slender body assumption. This was the first suitable theory for numerical computations of ship motions that had adequate accuracy for engineering applications [7]. A modified strip theory approach of Gerritsma and Beukelman [8] was shown to obtain a good agreement with experimental tests for head wave case. Ogilvie and Tuck [9] developed a mathematically consistent approach based on short wave-length approximation, conducting a systematic analysis for the slender body problem to determine the added mass and damping for heave and pitch motions. Most of today's strip methods are variations of the approach proposed by Salvesen et al. [7], which is one of the most complete versions, solving five degrees of freedom in ship's motion equations, as the surge component was neglected, being not relevant for a slender body oscillation. Since then, more comprehensive strip theories have been developed for ship design, such as [10–14]. Few comprehensive reviews of strip theory variations have been reported by [4,15,16]. Most recently, a combination of two-dimensional strip theory with the two-dimensional Green function based on the potential theory to solve boundary values and motion responses of a semi-planing craft have been reported by [17]. Despite theoretical shortcomings, strip theory has the advantage of being fast, cheap and sufficiently accurate for a range of hull forms and moderate speeds. However, for higher speed vessels, highly flared hull forms, wave loads or extreme motions, other approaches have to be used.

The effort devoted to overcoming the weakness of the strip theory methods and to modelling the non-linear phenomena leads to the development of other potential flow alternatives based on numerical 3D methods that enhance the modelling technique of the physical domain of interest. By the late 1970s, a new treatment of the boundary conditions led to the Neumann–Kelvin approach [2], which supposes that the body boundary condition is applied to the mean position of the exact body surface and linearized free-surface boundary condition [2]. The most efficient way to solve the Neumann–Kelvin problem is to employ boundary integral methods where the solution is formulated in terms of singularities, as sources or dipoles, over the hull and free surface [2]. Two different categories of methods have been developed based on the type of singularities used in integral equations, one relies on Green function [18], which automatically satisfies both the radiation and the linearized free surface condition, and the other one relies on the distribution of the simple Rankine sources [18], not only on the hull surface, but also on the free surface. Some of the weaknesses of strip theory and Green function approach can be overcome by using a 3D time-domain panel method based on Rankine sources, including the fully 3D effects of the flow and forward speeds effects.

Hess and Smith [19] proposed in 1964 a method for evaluating velocity and pressure fields around a fully immersed arbitrary three-dimensional body. The method, also called the panel method or the boundary element method allowed, for the first time, to calculate the flow around an arbitrary three-dimensional body imposing the boundary condition exactly on the body surface. Firstly Gadd [20], then Dawson [21], applied the panel method for steady free surface flow distributing Rankine sources on hull and undisturbed free surface. A wide variety of different approaches for solving the nonlinear steady forward motion problem, based on Rankine sources distribution, has been reported by [22–25]. Due to the development of computing capabilities, the panel method has been continuously improved to solve the problems of ship motions in waves considering different modelling of nonlinearities of the hull and free surface boundary conditions. Nakos et al. [26] used a Rankine source method to solve transient wave–body interactions. This method has linearized the solutions about the double body flow. Large ship motion analysis performed by [27] was based on the application of the desingularised source method. Söding et al. [28] applied patch method instead of the collocation method to satisfy boundary conditions on the solid body surface, considering for the free surface a nonlinear condition. Recently, an improved desingularised Rankine method has been proposed by Mei et al. [29] to solve 3D diffraction and radiation problems. Dai and Wu [30] combined a method based on a Rankine panel method for the near-field and a transient Green function method for the far-field to predict large amplitude ship motions.

Potential flow solvers for seakeeping prediction have been continuously improved in terms of efficiency and accuracy, but still, the effect of viscosity, turbulence, wave dispersion, wave breaking, green water and slamming impact loading, deck green sea cannot be captured properly using the method based on potential flow assumption [31,32]. Considering the steadily increasing computational power and parallel computing, computational fluid dynamics methods based on the Reynolds-averaged Navier–Stokes (RANS) approach is more frequently applied to solve unsteady seakeeping problems. Most of the commercial solvers adopt a combination of unsteady RANS method, based on finite volume method and multi-phase flow approach, using the volume of fluid method for the treatment of free surface, which is rapidly gaining popularity for ship's motions applications. The advantages of the nonlinear computation techniques without using analytical formulas for added resistance or empirical values for viscous effect are significant for the seakeeping analysis accuracy, but they are considerably more time consuming than potential flow approaches. Therefore, a good balance between accuracy and computational speed is required, especially for the ship design process.

The majority of RANS seakeeping simulations have been performed at model scale. The first attempt to solve ship motions in waves was presented in [33]. The results show some problems regarding the accuracy of free-surface due to the limited grid quality. Later, Simonsen et al. [34] carried out motions analysis in heave and pitch motions in regular head waves for KRISO container ship (KCS) hull using the CFDSHIP-IOWA code. Recently, Lungu and Bekhit [35,36] have assessed the seakeeping performances of the KCS and KVLCC (KRISO Very Large Crude Carrier) ship models for regular head wave conditions using NUMECA/FineMarine code. Their results show good agreement with the experiment for model scale. On the other hand, the scale effect seems to be important, as Hochkirch and Mallol [37] presented significant scale effects due to the differences between model-scale flows and full-scale flows. Tezdogan et al. [38] investigate seakeeping behavior and performance of the KCS model at a slow forward speed using Star-CCM+ package (Siemens PLM Software, Plano, TX, USA).

Considering the context described above, the present work is focused on a systematic comparison study concerning the analysis of ship's pitch, heave and roll motions in regular waves, for a full-scale ship model. The study was made using three different numerical approaches: the linear strip theory based on the Lewis form shape parameterization developed by Domnisoru [39], as in-house code DYN (module OSC - Oscillations), the non-linear potential flow based on Rankine source distribution using the commercial code SHIPFLOW and, for a limited number of cases, due to it being a highly time consuming solution, the viscous RANS method using NUMECA/FineMarine commercial code. Several series of computations have been performed for the regular wave height of 2 m, heading angle in the range of 0 to 180 deg, with a step of 45 deg. For the BEM model also, a wave height of 8 m has been considered. The computations have been carried out for the full-scale KCS hull numerical model, considering two forward speeds, 12 and, respectively, 24 knots.

The authors of this article aim to integrate all the presented methods in the same study in order to achieve a methodology that serves the rapid decision-making process in the case of further studies concerning the choice of the most accurate estimation method for ship motions on waves, depending on the specific requirements of the study and taking into account costs and calculation time. The proposed methodology meant approaching the mentioned methods, globally, with in-house code and commercial software, and the major advantage consisted of a better management of the input data, working conditions, calculation hypotheses, analyzed cases. It is worth mentioning that in the present article, full-scaled ship computations are considered, while most of the literature references reported analyses on model-scaled ship.

2. Methods for Ship Motions Prediction

As most of the available techniques to predict ship motions for design purposes rely on assumptions of potential flow theory, two different methods from this category have been chosen for comparison. The third approach considered for the present study is based on RANS solver, but, despite the parallel computing, this method does not prove yet to be able to solve the seakeeping problem in reasonable time to be consider in the practical design procedures. Taking into account that the methods involved in the present study rely on well-known mathematical formulation presented in seakeeping references [4,7,22,40–42], in this chapter only a brief presentation for each method is included.

The first method is based on a potential flow hydrodynamic linear 2D strip theory, with Lewis ship's stations parameterization and regular wave Airy model excitation, with a frequency-domain solution of the coupled heave–pitch motion equations and uncoupled roll motion equation. This method is implemented by Domnisoru [39]—as in-house code DYN, module OSC—for the linear oscillations response amplitude operator (RAO) response amplitude operator computation. The mathematical formulation of the method used is described in [39]. The DYN code, module OSC, has been validated by experimental tests on scaled models of a fishing vessel [43] and a survey vessel [44], on follow, head, beam and quartering regular wave conditions.

For the second set of computations, a 3D fully nonlinear time-domain boundary element method, implemented into the commercial code SHIPFLOW [22,40] has been employed for ship motion estimation, considering only potential flow formulation being suitable for practical use. The method solves the missed boundary value problem for Laplace equation by distributing sources on the hull and free surface. Integrating the kinematic and dynamic boundary conditions with a fourth order Adam–Bashford–Moulton method, the elevation of free surface is obtained at each time step based on a Mixed Euler–Lagrange method. A blending zone based on an analytical solution is introduced in order to avoid the reflection from domain boundaries and in order to generate a certain incident wave. Once the velocity potential is obtained on the hull, using Unsteady Bernoulli formulation, the pressure distribution on the hull can be obtained. Finally, after integrating pressure on the hull, forces, moment and then motions are determined. A full description of the mathematical model and the numerical method may be found in [40,45]. Validation of the above-described method has been reported by Larsson et al. [46] for added resistance, pitch and heave, and by Coslovich et al. [47] mainly for roll.

The commercial software NUMECA/FineMarine [41,42] has been used in this study to solve the incompressible unsteady RANSE flow at ship motions in head waves. The solver relies on the finite volume method to build the spatial discretization for the governing equation. Closure to the turbulence is achieved by making use of the k-ω SST model with wall function formulation [41]. Pressure–velocity coupling is enforced through a SIMPLE (Semi-Implicit Method for Pressure Linked Equations)-like approach, where the velocity updates come from the momentum equation and the pressure is extracted from the mass conservation constraints transformed into pressure equation. Convection and diffusion terms in the RANSE are discretized using a second-order up-wind scheme and a central difference scheme, respectively. Free-surface capturing strategy is based on multi-phase flow approach using a volume of fluid method with high-resolution interface schemes [42]. Ship's 6 DOF (Degrees Of Freedom) can be solved by the solver, but also some degree of freedom can be restrained. Mathematical model and the numerical framework are nicely described in [41]. Several validation studies of the solver have been reported by [48] for ship resistance and self-propulsion, and by [31,35,36,49] for ship motions in head waves. Their results show good agreement with the experiment for the scaled model.

In the preliminary ship design process, time is an important aspect to be considered due to the fact that naval architect usually cannot afford too long computational time. Considering the present methods comparison for one case (v = 12 Kn, ω = 0.4, h_w = 2 m and head wave), by far strip theory is the less time-consuming method, the time spend for a single calculation case was 3 min using a single processor. The nonlinear BEM computations have been performed on 4 processor machines and the physical computational time for the same case was 19 h and 5 min for a number of about 68,000 panels. The RANS computations have been carried out on a high-performance computing (HPC) machine using 120 processors for a grid of about 20 million cells, which led to a 41 h and 47 min computational time. On the other hand, time consuming should be also correlated with the accuracy and with the amount of information received from a specific flow post-processing. The RAO functions for heave, pitch and roll based on the linear strip theory implemented in DYN code have been validated in references [43,44] for heading angles, 0, 90, 180 degrees, proving good results for practical investigations, revealing average differences from 12% to 18%. A validation BEM method in terms of heave and pitch was performed by the authors in [50] and the computed results showed good agreement of 3% to 10% with towing tank measurements. As mentioned before, extensive validation studies may be found in [45,46]. In addition, grid convergence and validations studies based on FineMarine RANS solver have been reported by Lungu and Bekhit [35,36]. Their results revealed good agreement with the towing tank test measurements for model scale, 1–7%. Taking all the above mentioned into consideration, a trade-off between accuracy and computational costs must be counted in order to obtain feasible results for the ship design practical use.

In order to compare directly the computational fluid dynamics (CFD) results with those obtained by using linear strip theory, for all numerical analyses, the regular Airy model is considered as excitation source. Both CFD methods are hydrodynamic nonlinear, so that even if the excitation has one harmonic, the time-domain response has several harmonics. For the two CFD methods, heave, pitch and roll time records are spectrally analyzed by using Direct Fourier Transformation Method. Regarding the response amplitude spectra (see Section 6), only the motions amplitudes with encountering wave frequency harmonic are considered. The response amplitude operators (RAO s) for ship motions obtained by the two CFD methods represent the ratio between heave, pitch and roll amplitudes on wave's harmonic and the regular wave amplitude.

3. Hull Geometry and Conditions

The objective of the present study is to investigate the seakeeping performances of the KRISO container ship (KCS) while sailing in regular wave at two speeds, 12 and 24 Kn, corresponding to

Froude numbers of 0.13, and 0.26, respectively. The KCS hull is a benchmark test case for ship hydrodynamics widely used in the marine hydrodynamic scientific community [34]. The hull is a modern commercial container vessel with bulbous bow and flare fore above the waterline. The stern is a typical pram type stern with transom. Bare hull, without rudder, has been considered for the seakeeping calculations. KCS hull geometry is presented in Figure 1 and the main dimensions and conditions used for numerical simulations are presented in Table 1.

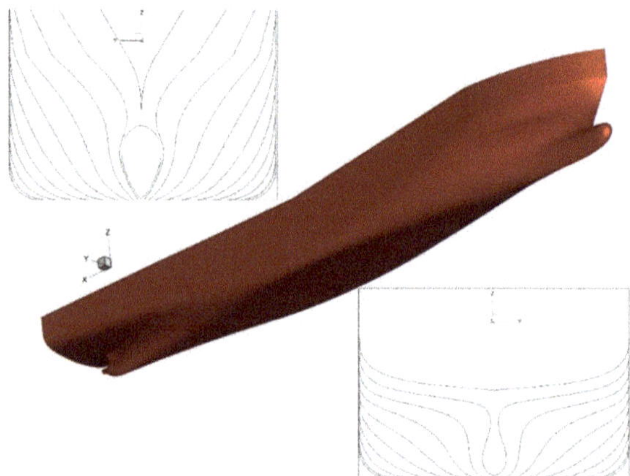

Figure 1. KRISO container ship (KCS) hull geometry.

Table 1. KCS ship main dimensions.

Dimension	Value	Units
Length between the perpendiculars (L_{BP})	230	m
Length of waterline (L_{WL})	232.5	m
Beam at waterline (B_{WL})	32.2	m
Depth (D)	19.0	m
Design draft (T)	10.8	m
Displacement (Δ)	52,030	m^3
Block coefficient (C_B)	0.6505	-
Ship wetted area (S)	9422	m^2
Longitudinal center of buoyancy (L_{CB}) (%L_{BP}), fwd+	-1.48	
Longitudinal center of gravity (L_{CG}) from the aft peak	111.603	m
Vertical center of gravity (KG) from keel	10.8	m
Moment of inertia (Kxx/B)	0.35	-
Moment of inertia (Kyy/L_{BP}, Kzz/L_{BP})	0.25	-
Speeds (v)	12.24	Kn

In Table 2, the test cases for seakeeping analysis are presented. The ship motions are approached considering the ship on regular Airy wave, having the wave height of 2 and 8 m, the wave frequency between 0 and 2 rad/s and heading angle in range of 0 to 180 deg, with a step of 45 deg. Even though the ship response in a regular wave at zero speed was computed, in the following paragraphs the ship motions for only two speed values of 12 and 24 Kn, respectively, are presented. Although, the study was carried out for three values of vertical center of gravity, $KG = 7.3$, 10.8 and 14.3 m, in the paper, the results for only $KG = 10.8$ m, which corresponds to a medium loading case of the ship, are presented.

Table 2. Test cases.

h_w (m)	v (Kn)	Method	μ (deg)	ω (rad/s)											
2	12	ST	0, 45, 90, 135, 180	0					Step 0.01						2
		BEM	180	0.3	0.35	0.4	0.45	0.5	0.55	0.6	0.65	0.7	0.8	0.9	1
		RANS	180	0.3		0.4	-	0.5	-	0.6	-	0.7	-	-	-
	24	ST	0, 45, 90, 135, 180	0					Step 0.01						2
		BEM	180	0.3	0.35	0.4	0.45	0.5	0.55	0.6	0.65	0.7	0.8	0.9	1
8	12	ST	0, 45, 90, 135, 180	0					Step 0.01						2
		BEM	180	0.3	0.35	0.4	0.45	0.5	0.55	0.6	0.65	0.7	0.8	0.9	1
	24	ST	0, 45, 90, 135, 180	0					Step 0.01						2
		BEM	180	0.3	0.35	0.4	0.45	0.5	0.55	0.6	0.65	0.7	0.8	0.9	1

4. Numerical Setup

Several details about the computational conditions will be briefly described in the following. For the discretization of the hull used for the linear 2D strip theory calculations, a model of 526 stations has been used.

For the BEM method, the boundary value problem modelling the field equation and boundary conditions have been solved, making use of the commercial code SHIPFLOW. The computations carried out with BEM is a fully nonlinear unsteady three-dimensional potential flow method, that takes into account the geometric non-linearities of ship shape and the induced hydrodynamic non-linearities, although the external excitation is kept linear. Wave reflection from domain's boundaries is a pure numerical problem that affects all the simulations with a restricted domain. In order to avoid wave reflection from the free surface truncation boundaries, the flow on the intersection between the outer and inner domain has to be the same. To achieve this condition, a numerical damping zone is located close to the domain's boundaries and such a zone only has to dampen the difference between the velocity potential in the inner and outer domain. This difference in the solution between the inner and the outer domain is mainly due to the presence of the body with radiation and diffraction effects but can also derive from a numerical error. This damping is achieved adding a term in the free surface boundary condition for all those panels that belong to such a zone [51,52]. The fluid domain assumes that the flow is known a priori in the outer domain and the current method uses analytically described waves to represent the flow field in the outer domain. The hull and free surface have been discretized by panels. The calculations have been performed for a surface domain of one ship length upstream, two ship length downstream, the width of the free surface being two ship length. For BEM method, the ship hull is discretized both sides without the request of symmetry condition, so that any heading angle condition can be computed. The number of panels distributed on hull and free surface varied between 60,000 and 90,000 panels function of the wave frequency case studied, which assures a minimum number of 40 panels per wavelength. The computations have been performed on local desktop machines with four cores of 3.1 GHz.

For the unsteady RANS simulation of incompressible flow of ship motions in head waves, the commercial solver NUMECA/FineMarine was used. A reference length has been considered $L_{ref} = \max (L_{BP}, \lambda)$. The dimensions of the computation domain have been chosen according to the ship and wavelength, considering 2.0 L_{ref} upstream, 4.0 L_{ref} downstream, 2.0 L_{BP} on the side, 4.0 L_{BP} underneath, and 2.0 L_{BP} above the undisturbed free-surface level. The boundary conditions imposed on the solid wall and domain boundaries, but also the dimensions considered for the computational domain, are depicted in Figure 2. A mirror condition is applied on the centerline plane of the ship and on the side boundary of the domain, to avoid wave reflections at lateral boundary. Considering the head wave case computed by the RANS method, the corresponding wave is generated at the upstream boundary and the inlet boundary is chosen to ensure at least two full waves to be generated before encountering the ship. The free surface is refined for the entire domain using three different refinement boxes, as can be observed in Figure 2. The first starts from the inlet boundary and is extended until it reaches 1.0 L_{ref} behind the ship. The cell size is chosen for this refinement zone in x- and y-direction to provide minimum 60 cells per wavelength, such that $\Delta x1 = \Delta y1 = \lambda/60$, while in the z-direction the refinement criterion is selected as $\Delta z = h_w/16$, where h_w represents the wave height. The refinement depth in z-direction is extended for 3.0 h_w equally distributed above and beneath the non-disturbed free-surface level, which is set at the design draft $T = 10.8$ m from the ship base line. The second and third boxes are coarsened gradually in x- and y-directions to generate sufficient numerical damping zones to prevent any reflections from the exit boundary. The second box is extended for 1.0 L_{ref} and it is used as an initial damping relaxation zone, in which the cell size is coarsened in x- and y-directions by a factor of 4 (i.e., $\Delta x2 = \Delta y2 = 4.0 \times \Delta x1$). The third refinement zone is used as a final damping zone, having the cell size coarsened in x- and y-directions by a factor of 8.0 (i.e., $\Delta x3 = \Delta y3 = 8.0 \Delta x1$), while the refinement in the z-direction for the entire domain is maintained unchanged. The effect of damping zone on the RANS solution is clearly seen in Figure 3.

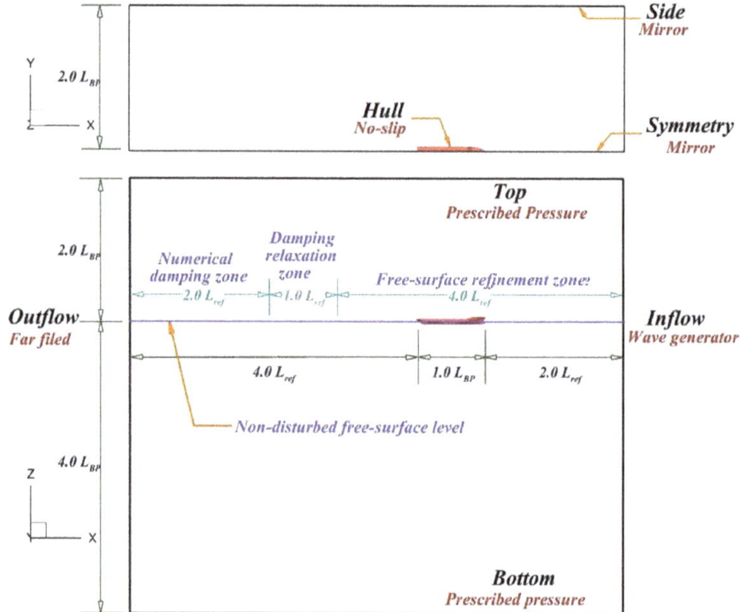

Figure 2. RANS Computational domain, dimensions and boundary conditions.

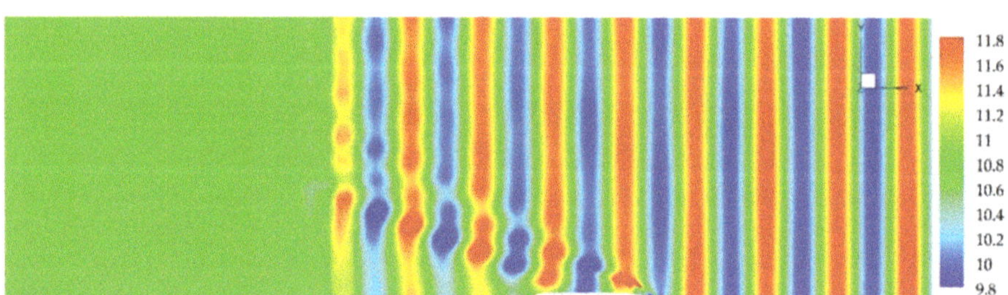

Figure 3. RANS damping zone effect.

Body-fitted full hexahedral unstructured meshes of about 22 million cells have been generated (Figure 4). The time step Δt has been chosen in order to fulfil the condition of 300-time steps per wave period, with fourth order convergence criteria. All the computations have been performed on a HPC machine with 120 cores of 3.3 GHz.

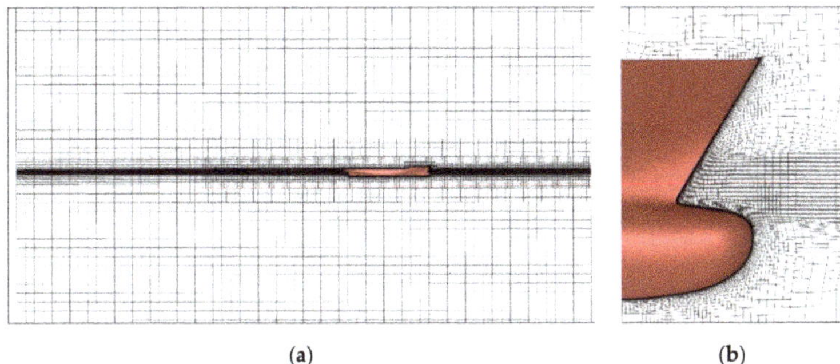

Figure 4. Computational grid. (**a**) Domain discretization; (**b**) mesh details in bow area.

In general, methodology used in the present paper for the seakeeping performance prediction is based on the practical guidelines recommended by the International Towing Tank Conference (ITTC), by software providers and different sensitivity and validation studies [35,36,49–52], but for some particular cases the recommendations could not be accomplished due to the practical reasons.

5. Grid Convergence Test

A grid study for SHIPFLOW calculations was conducted for three grids based on the Richardson extrapolation [53] in order to investigate the numerical simulation error and uncertainty. Two calculation cases have been considered for the grid test: ship on regular waves, head waves (μ = 180 deg) and beam sea (μ = 180 deg). For both cases ship speed was v = 12 Kn. The results of grid convergence study for RAO heave, pitch and roll are summarized in Table 3. The convergence ratio is defined as $R_G = \varepsilon_{21}/\varepsilon_{32}$, where $\varepsilon_{21} = S_2 - S_1$ and $\varepsilon_{32} = S_3 - S_2$ stands for simulation error, S_1 represents the solution calculated for fine grid (68,716 panels), S_2 for medium grid (32,692 panels) and S_3 for coarse grid (16,346 panels), according to Richardson extrapolation approach [53], with the expressions (1) ÷ (4). The grid corresponding to each level of refinement is depicted in Figure 5 for hull and free surface.

Table 3. Grid convergence test results.

Variable	μ = 90 deg		μ = 180 deg	
	Heave	Roll	Heave	Pitch
S_1 (fine)	1.101188	0.031396	0.713147	0.013927
S_2 (medium)	1.101366	0.031428	0.714934	0.013929
S_3 (coarse)	1.104284	0.031683	0.719730	0.014070
r_G	2.1	2.1	2.1	2.1
ε_{21}	0.000179	0.000032	0.001788	0.000001
ε_{32}	0.002917	0.000254	0.004796	0.000141
R_G	0.061195	0.127175	0.372726	0.010014
p_G	3.7654	2.779463	1.330184	6.109572
δ_G	0.000012	0.000005	0.001062	1.432×10^{-8}
U_G	0.000093	0.000014	0.001062	7.913×10^{-7}

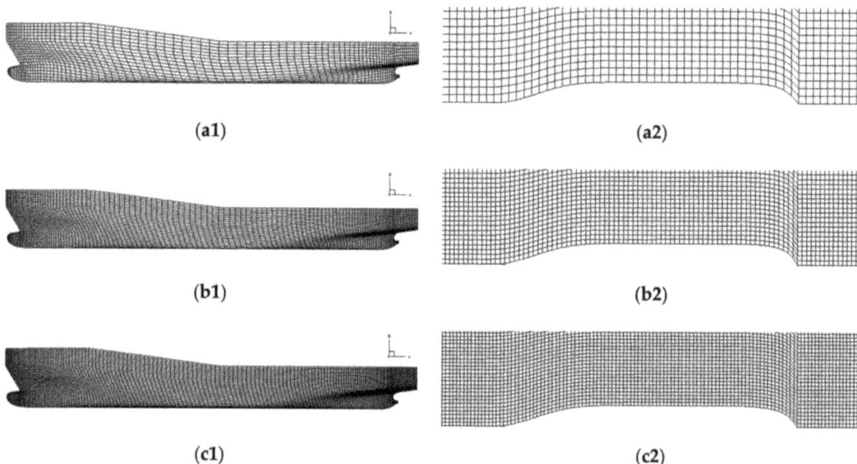

Figure 5. Grid refinement levels used for the grid convergence test. (**a1**) Coarse grid on hull surface; (**a2**) Coarse grid on free-surface; (**b1**) Medium grid on hull surface; (**b2**) Medium grid on free-surface; (**c1**) Fine grid on hull surface; (**c2**) Fine grid on free-surface.

Considering the computed values of R_G for all variables of grid convergence study, one can observe that monotonic convergence is achieved for all the simulation variable. The order of accuracy p_G can be calculated based on the refinement ration r_G, as follows:

$$p_G = \frac{\ln\left(\frac{\varepsilon_{32}}{\varepsilon_{21}}\right)}{\ln(r_G)}. \tag{1}$$

The calculated order of accuracy is used further to determine the correction factor C_G and the error δ_G, as follows:

$$C_G = \frac{r_G^{p_G} - 1}{r_G^{p_{th}} - 1}, \tag{2}$$

$$\delta_G = \frac{\varepsilon_{21}}{r_G^{p_G}}, \tag{3}$$

where p_{th} is the theoretical grid convergence order usually considered $p_{th} = 2$. Finally, the uncertainty is computed by the expression:

$$U_G = |C_G \times \delta_G| + |(1 - C_G) \times \delta_G| \tag{4}$$

Based on this study, bearing in mind that the percentage estimated error with respect to the fine grid is very small, i.e., $\varepsilon_{21}\%S_1 \ll 1$, which concludes that the solution computed by SHIPFLOW is grid independent.

6. Results and Discussion

In this section, the ship hydrodynamic response in regular waves is presented and discussed. The results are represented in the frequency domain, in terms of response amplitude operators (RAO) for heave, pitch and roll motion, respectively. The charts contain the results obtained by linear strip theory method (DYN code) and the boundary element method (SHIPFLOW software), also including a few results obtained using the RANS method (NUMECA/FineMarine).

In Figure 6, the ship on regular waves, for the case of top speed v = 12 Kn, μ = 90 deg, ω = 0.4 rad/s and h_w = 2 m, is presented. The images represent the hull position at five different fractions of ship motion periods (ΔT) related to free surface topologies. Hydrodynamic forces and moments are determined by integrating pressure over the ship hull based on the quadratic pressure distribution. Pictures on the left, Figure 6a1–e1, present the pressure distribution over the ship hull. One may see that the flow solution for ω = 0.4 rad/s reveals that the wave is fully developed, and artificial damping condition imposed on the boundaries works fine. Moreover, the interference of the wave system generated by the ship and the incident wave is well captured by the boundary element method and some nonlinear effects are expected (see the pictures on the right, Figure 6a2–e2).

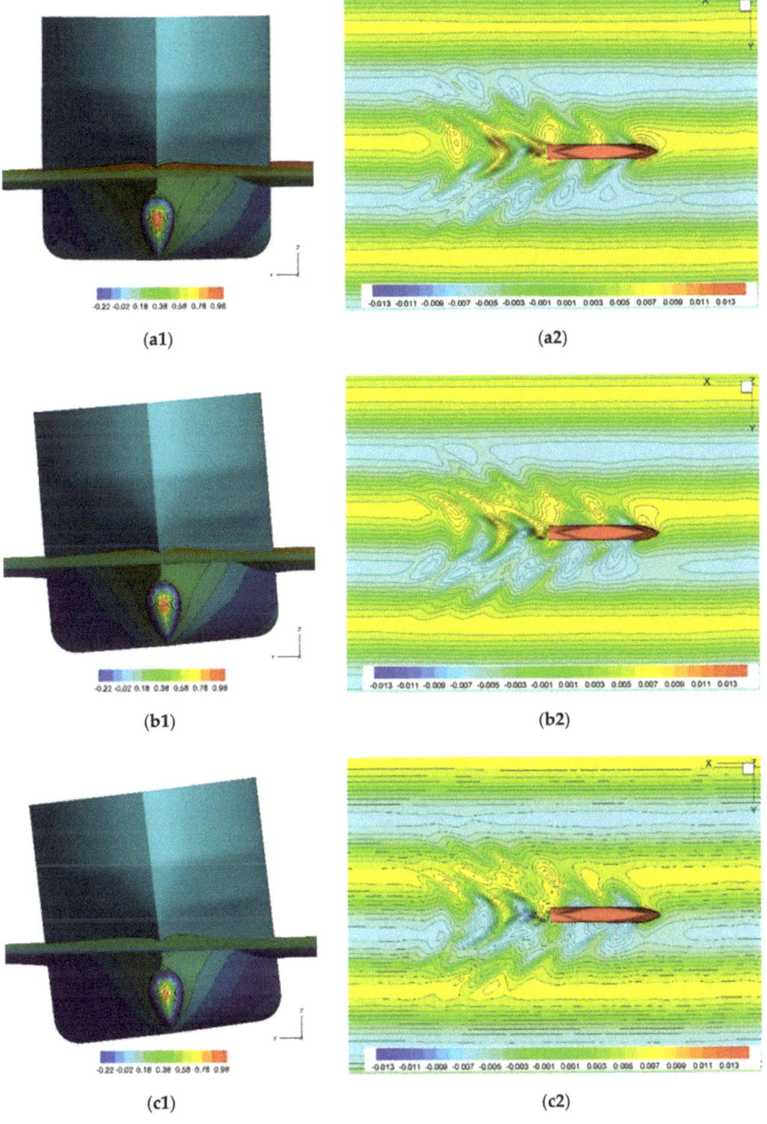

Figure 6. *Cont.*

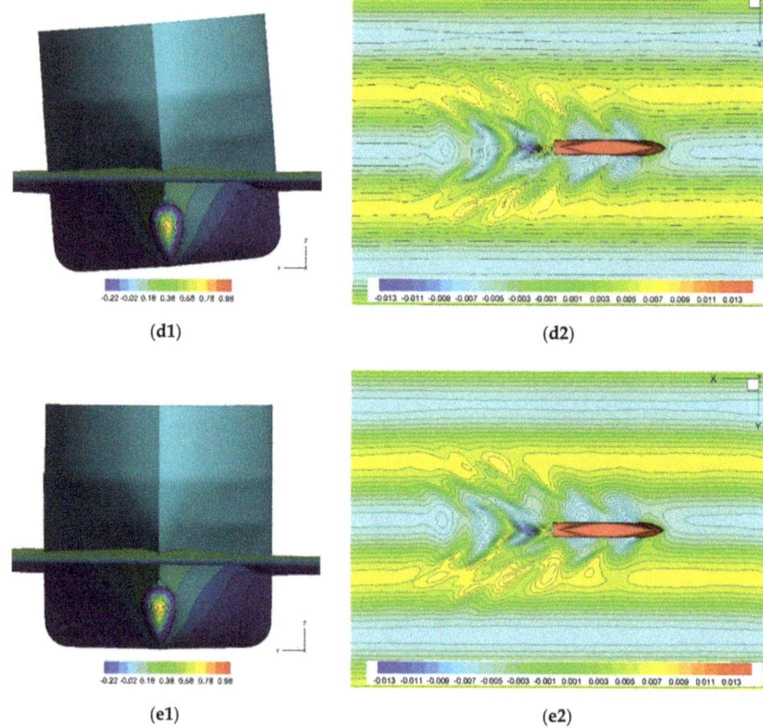

Figure 6. Pressure distribution and free surface topology computed for the case $\mu = 90$ deg. (**a1**) Pressure distribution on the hull, $\Delta T = 0$; (**a2**) Free-surface topology, $\Delta T = 0$; (**b1**) Pressure distribution on the hull, $\Delta T = 0.125$; (**b2**) Free-surface topology, $\Delta T = 0.125$; (**c1**) Pressure distribution on the hull, $\Delta T = 0.250$; (**c2**) Free-surface topology, $\Delta T = 0.250$; (**d1**) Pressure distribution on the hull, $\Delta T = 0.375$; (**d2**) Fre-surface topology, $\Delta T = 0.375$; (**e1**) Pressure distribution on the hull, $\Delta T = 0.500$; (**e2**) Free-surface topology, $\Delta T = 0.500$.

In Figures 7–11, the response amplitude operators for heave, pitch and roll motions of KCS ship advancing in regular waves are represented. Heave RAO (m/m), Pitch RAO (rad/m) and Roll RAO (rad/m), respectively, for the speed case of ship speed 12 Kn and the wave height 2 m (amplitude $a_w = 1$ m). In Figure 7, the heave response amplitude operator, Heave RAO (m/m), for the speed case of 12 Kn, obtained using a 2D strip and BEM methods is represented. As a general remark, for both methods, the ship response is significant in the lower frequency domain, while, as the frequency increases, the ship response decreases considerably. The maximum values are recorded for the heading angle $\mu = 90$ deg, at $\omega = 0.756$ rad/s. Once the circular frequency of the regular wave increases, one may observe that both methods are in good agreement.

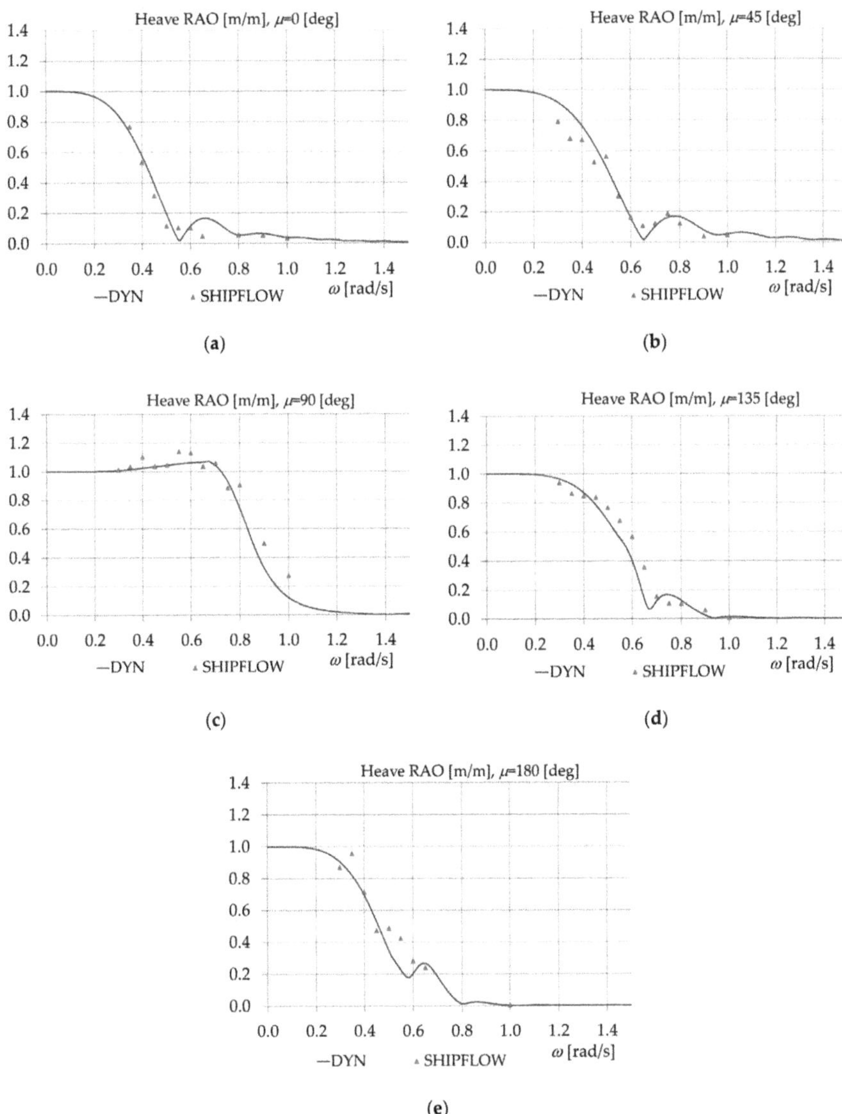

Figure 7. Heave response amplitude operator (RAO) (m/m) for $v = 12$ Kn, $h_w = 2$ m. (**a**) $\mu = 0$ deg; (**b**) $\mu = 45$ deg; (**c**) $\mu = 90$ deg; (**d**) $\mu = 135$ deg; (**e**) $\mu = 180$ deg.

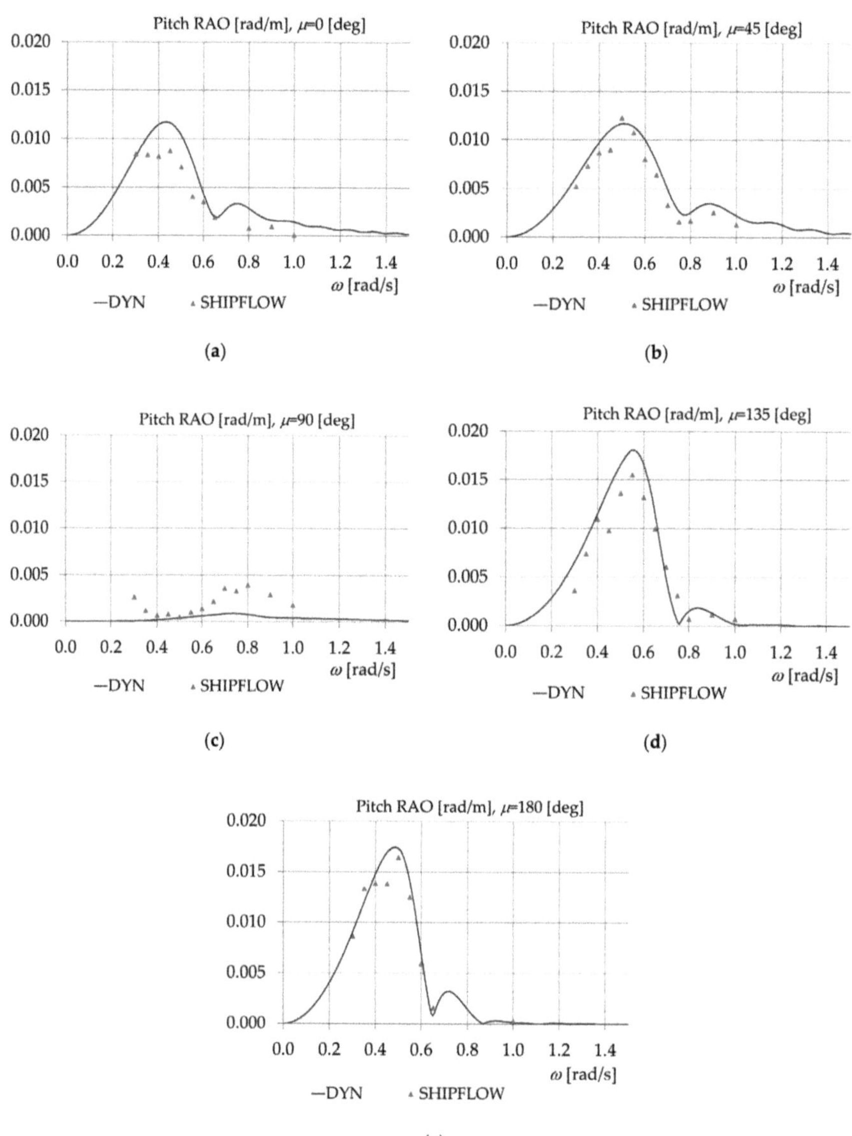

Figure 8. Pitch RAO (rad/m), for $v = 12$ Kn, $h_w = 2$ m. (**a**) $\mu = 0$ deg; (**b**) $\mu = 45$ deg; (**c**) $\mu = 90$ deg; (**d**) $\mu = 135$ deg; (**e**) $\mu = 180$ deg.

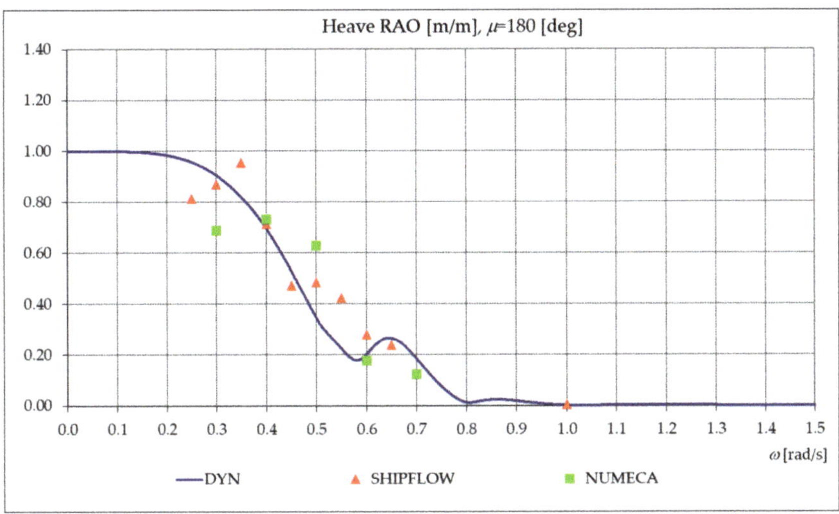

Figure 9. Heave RAO (m/m), DYN–SHIPFLOW–NUMECA comparison, $v = 12$ Kn, $\mu = 180$ deg, $h_w = 2$ m.

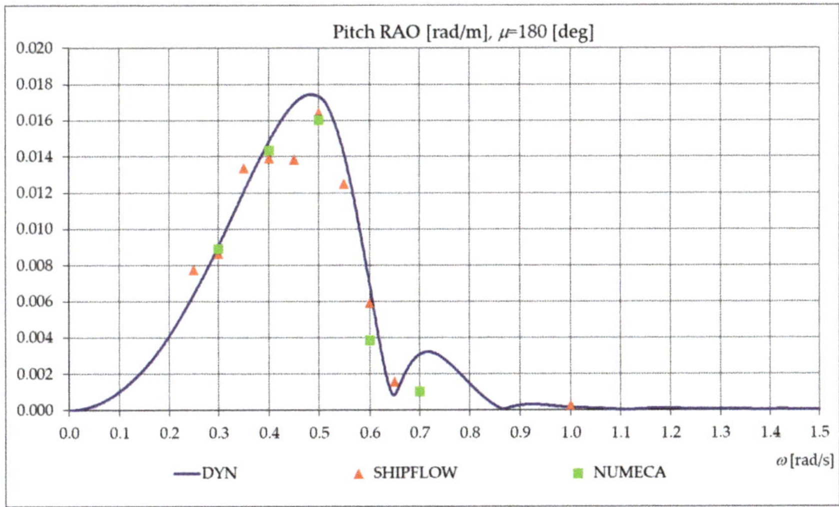

Figure 10. Pitch RAO (rad/m), DYN–SHIPFLOW–NUMECA comparison, $v = 12$ Kn, $\mu = 180$ deg, $h_w = 2$ m.

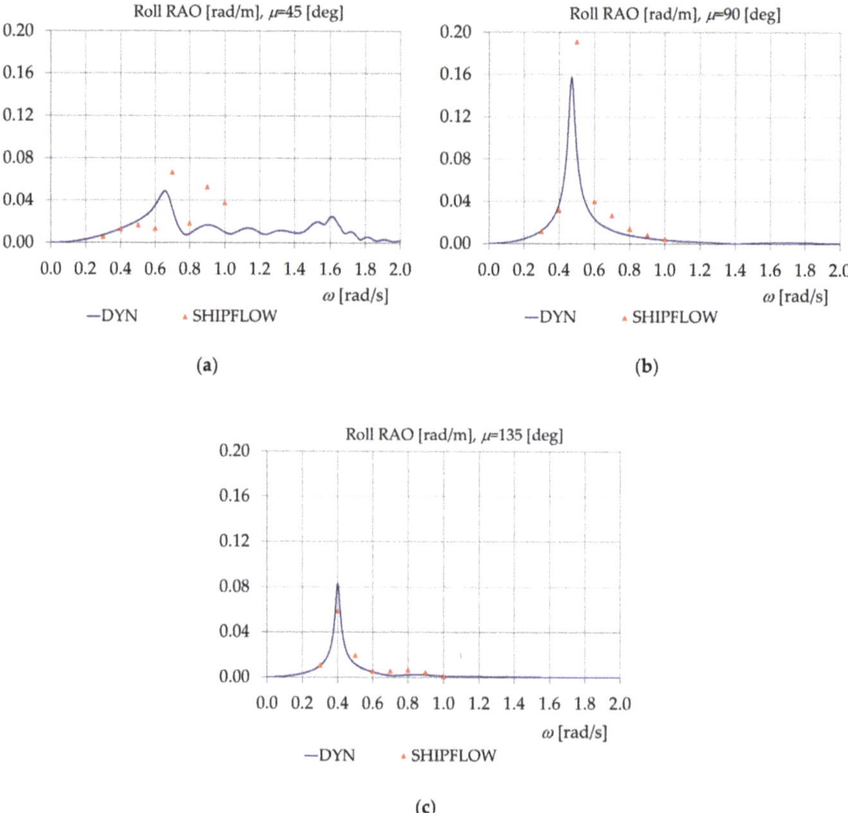

Figure 11. Roll RAO (rad/m), v = 12 Kn, KG = 10.8 m, h_w = 2 m. (**a**) μ = 45 deg; (**b**) μ = 90 deg; (**c**) μ = 135 deg.

In Figure 8, the response amplitude operator for pitch motion, Pitch RAO (rad/m) is depicted, for ship speed 12 Kn, heading angles from 0 to 180 deg, and wave height of 2 m. For pitch motion at beam sea condition, the response amplitude is the smallest one, recording scattered values between the 2D strip and the BEM hydrodynamic models. In particular, at 90 deg, the variation of BEM results has two peaks, the second one at similar frequency as the DYN results; however, the values obtained by the boundary element method are higher than the strip model results at beam sea. The maximum value is recorded at μ = 135 deg for ω = 0.57 rad/s. A good agreement between both hydrodynamic models is found at a heading angle of 45 and 135 deg.

In Figures 9 and 10, the following are presented: numerical results for Heave RAO (m/m) and Pitch RAO (rad/m) obtained with linear 2D strip method (DYN), BEM boundary element method (SHIPFLOW) and RANS method (NUMECA/FineMarine), considering the ship on head regular wave, μ = 180 deg, with 2 m height and a ship speed of 12 Kn.

In Figure 11, the following are given: the results for the roll motion in terms of response amplitude operator, Roll RAO (rad/m). The results are represented for the heading angle of 45, 90 and 135 deg, respectively. The maximum value for Roll RAO is recorded for μ = 90 deg and, as long as it moves away from the beam waves, the roll motion of the ship attenuates.

Next, a comparison between the results computed with linear 2D strip and BEM hydrodynamic models, for a regular wave height h_w = 8 m (a_w = 4 m) for BEM model, is depicted in Figures 12 and 13. First, one may see the Heave RAO (m/m) values at five different heading angles (Figure 12). The trend of the differences between the two methods remains the same as for the case of wave height $h_w = 2$ m. Very reduced differences occur in the case of the BEM method results at 2 and 8 m wave height, due to the reduced hydrodynamic nonlinearities. Higher differences appear at beam wave condition between 0.4 and 0.6 rad/s.

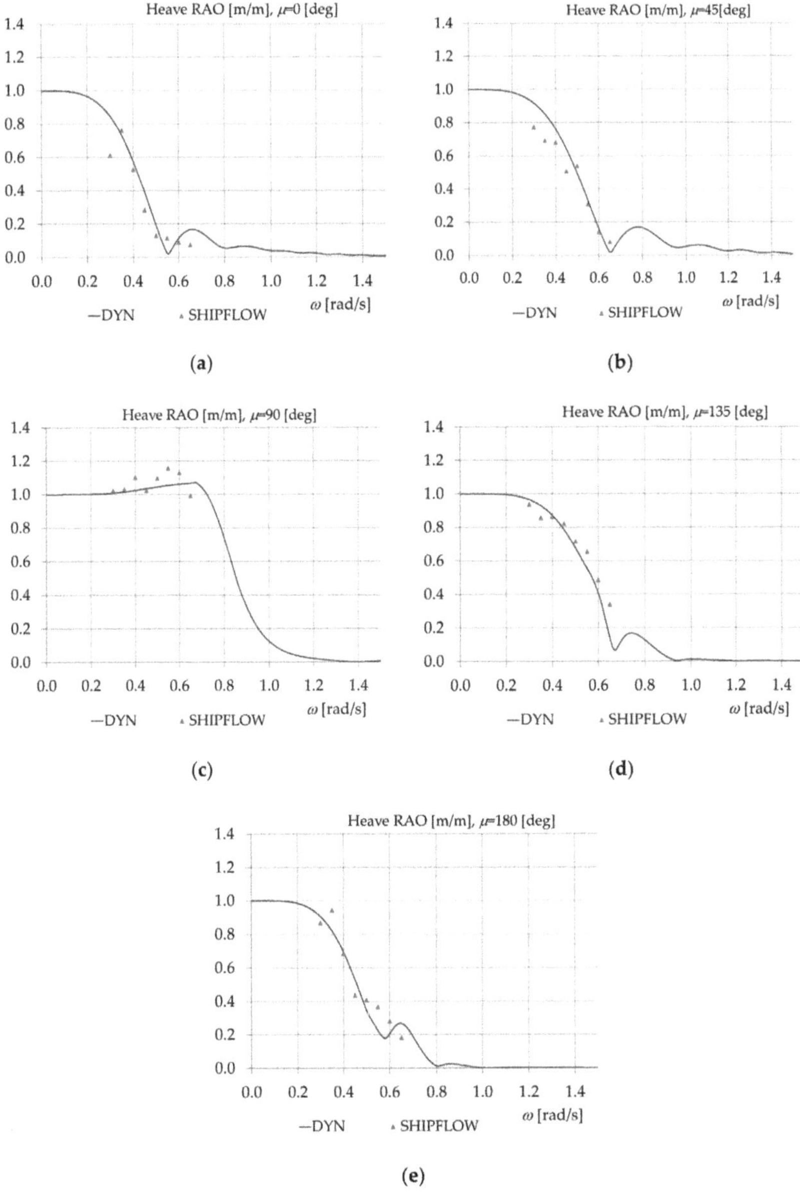

Figure 12. Heave RAO (m/m), $v = 12$ Kn, $h_w = 8$ m for BEM model. (**a**) $\mu = 0$ deg; (**b**) $\mu = 45$ deg; (**c**) $\mu = 90$ deg; (**d**) $\mu = 135$ deg; (**e**) $\mu = 180$ deg.

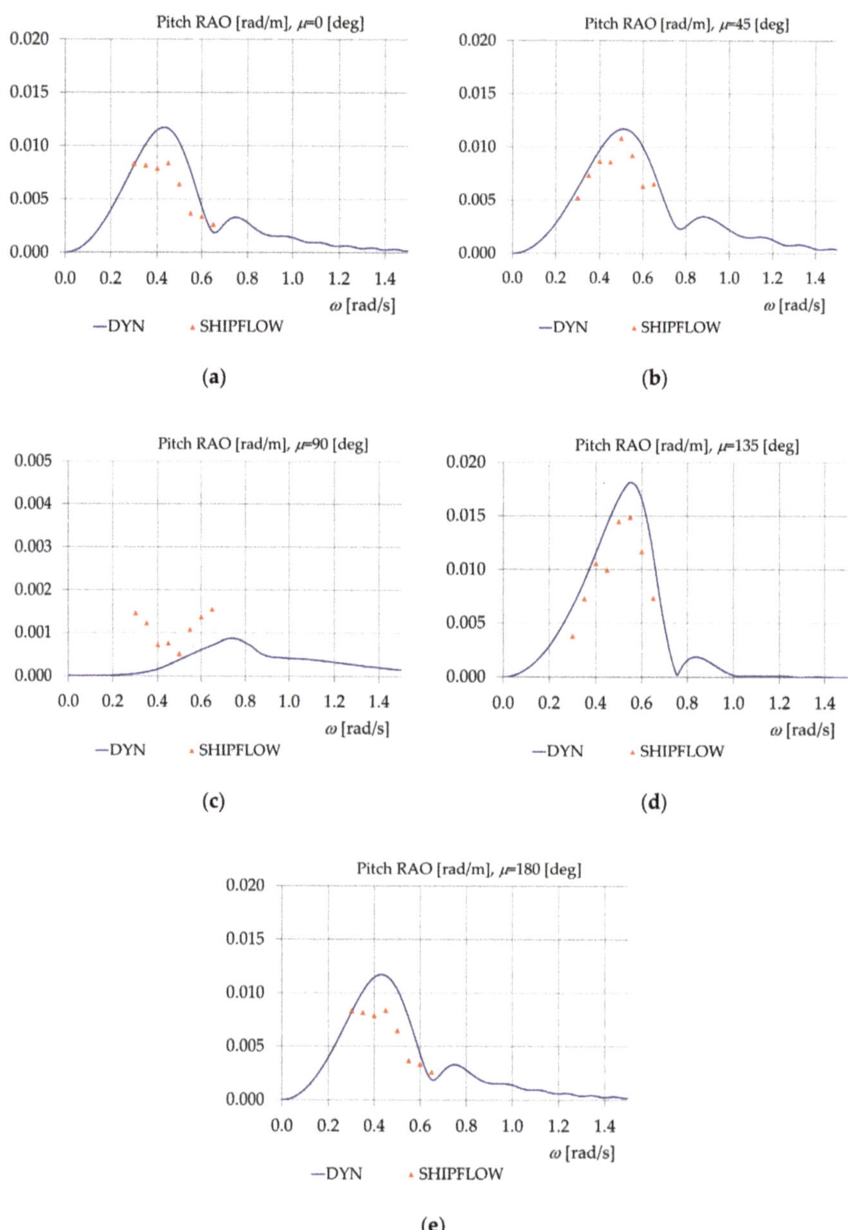

Figure 13. Pitch RAO (rad/m), v = 12 Kn, h_w = 8 m for BEM model. (**a**) μ = 0 deg; (**b**) μ = 45 deg; (**c**) μ = 90 deg; (**d**) μ = 135 deg; (**e**) μ = 180 deg.

In Figure 13, the variation curves of Pitch RAO (rad/m) for wave height h_w = 8 m for BEM model, with the peak value for the heading angle of μ = 135 deg around ω = 0.5 rad/s, are depicted. At beam waves, μ = 90 deg, differences between the hydrodynamic approaches are scattered, where the values of Pitch RAO are very reduced.

The next results are computed for the second ship speed case, v = 24 Kn, for both wave heights of 2 and 8 m, respectively. There were also considered the same medium loading conditions of the ship. In Figure 14, the RAO heave (m/m) is presented. The differences between the linear 2D strip and BEM hydrodynamic approaches have slightly increased compared to 12 Kn ship speed, especially for wave circular frequency lower than ω = 0.4 rad/s.

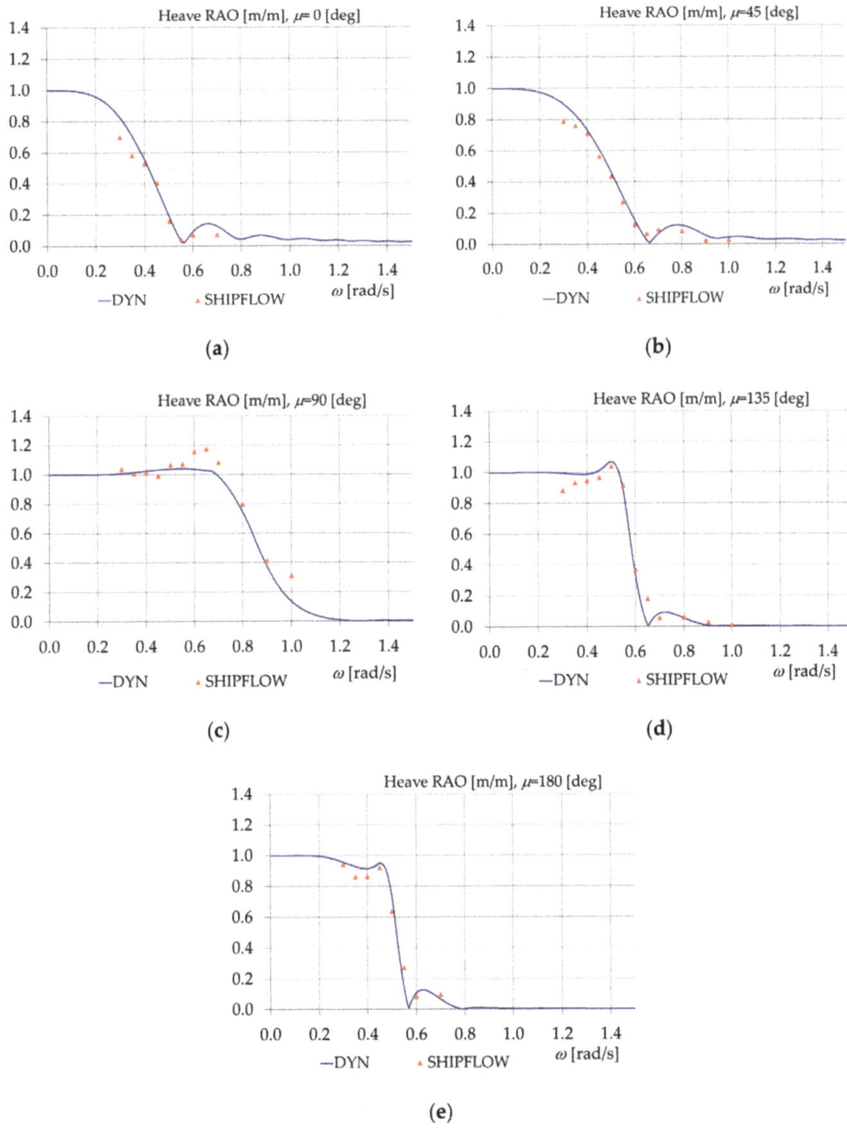

Figure 14. Heave RAO (m/m), v = 24 Kn, h_w = 2 m for BEM model. (**a**) μ = 0 deg; (**b**) μ = 45 deg; (**c**) μ = 90 deg; (**d**) μ = 135 deg; (**e**) μ = 180 deg.

In Figure 15, the Pitch RAO (rad/m) for 24 Kn ship speed and 2 m wave height are presented. The Pitch RAO has the maximum peak values for the head wave case, around ω = 0.46 rad/s. At the same time, one may observe that the hydrodynamic models provide better agreement for the beam wave case $\mu = 90$ deg.

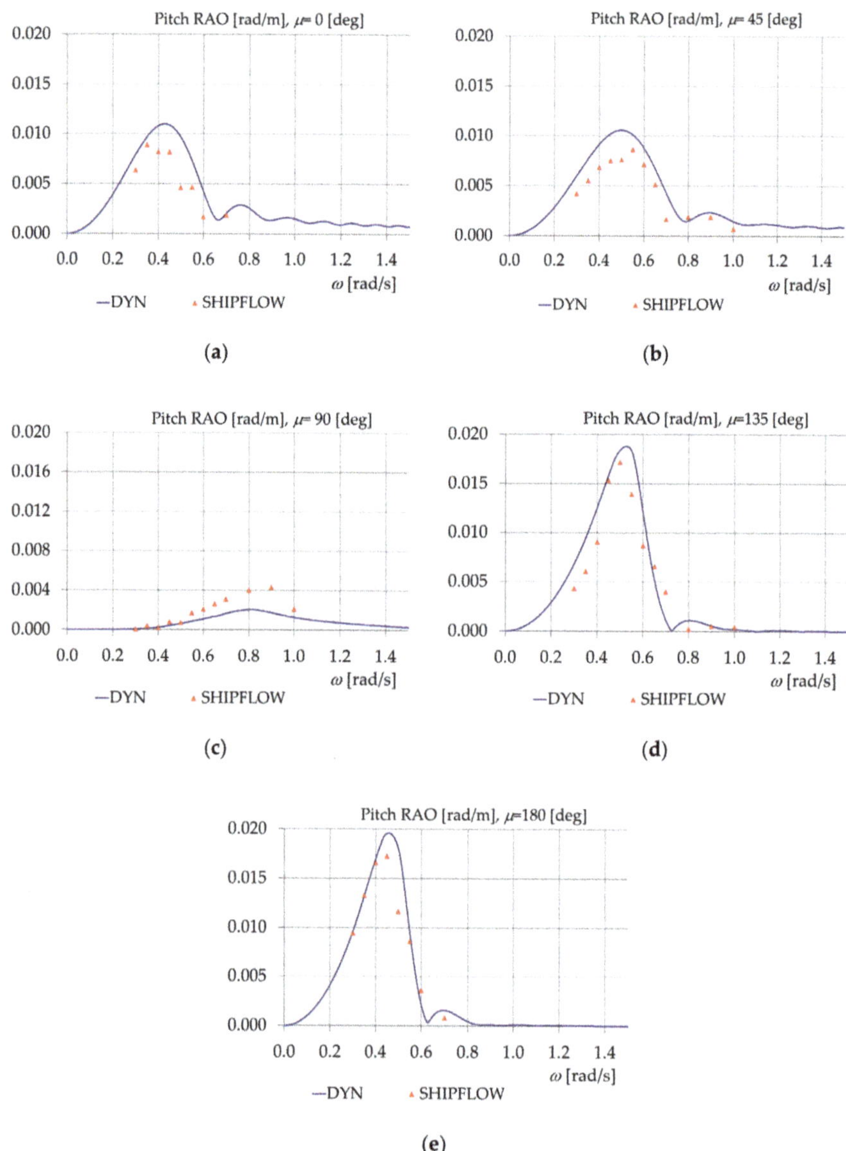

Figure 15. Pitch RAO (rad/m), $v = 24$ Kn, $h_w = 2$ m. (**a**) $\mu = 0$ deg; (**b**) $\mu = 45$ deg; (**c**) $\mu = 90$ deg; (**d**) $\mu = 135$ deg; (**e**) $\mu = 180$ deg.

In Figure 16, the Roll RAO (rad/m) for 24 Kn ship speed and 2 m wave height are presented. The trend of results is maintained for the roll motion as for the 12 Kn ship speed case, the boundary element method being in agreement with the strip theory, for all the heading waves considered.

The maximum values appear at beam wave, 0.16 rad/m around $\omega = 0.47$ rad/s. At the same time, as long as the ship is in oblique waves, the values of roll motion decrease.

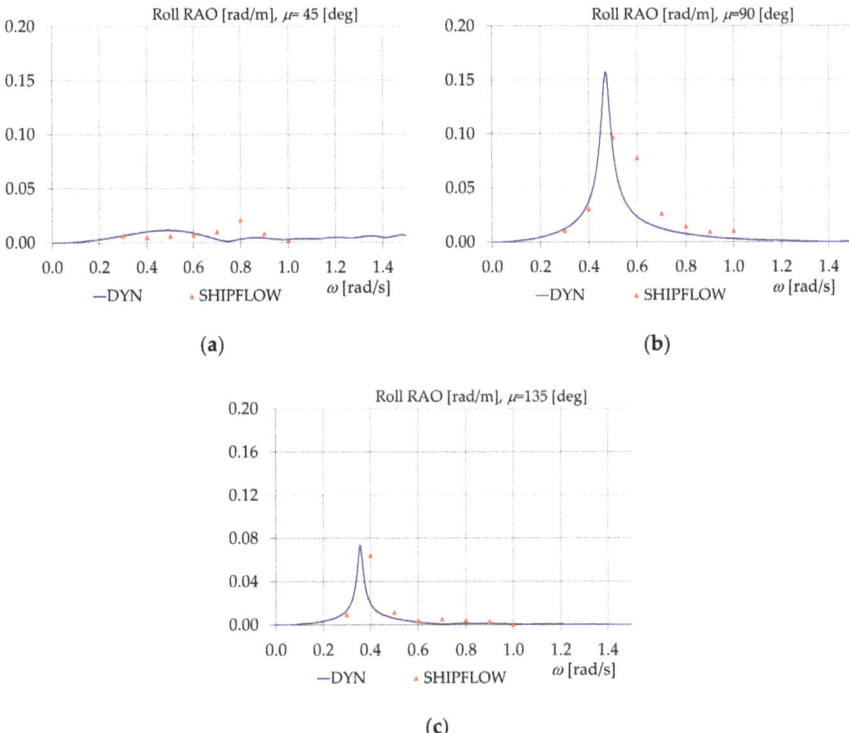

Figure 16. Roll RAO (rad/m), $v = 24$ Kn, $KG = 10.8$ m, $h_w = 2$ m. (**a**) $\mu = 45$ deg; (**b**) $\mu = 90$ deg; (**c**) $\mu = 135$ deg.

At 24 Kn ship speed and 8 m wave height for the BEM model, the Heave RAO (Figure 17) records visible differences to the BEM solution at the same speed, but with 2 m wave height, due to hydrodynamic nonlinearities for the wave's circular frequency up to 0.6 rad/s. Figure 18 presents the pitch RAOs for 24 Kn ship speed and 8 m wave height. The differences between the BEM model, with 8 m wave height, and the linear 2D strip model become more significant in comparison to the case with wave height 2 m, due to the recorded hydrodynamic nonlinearities.

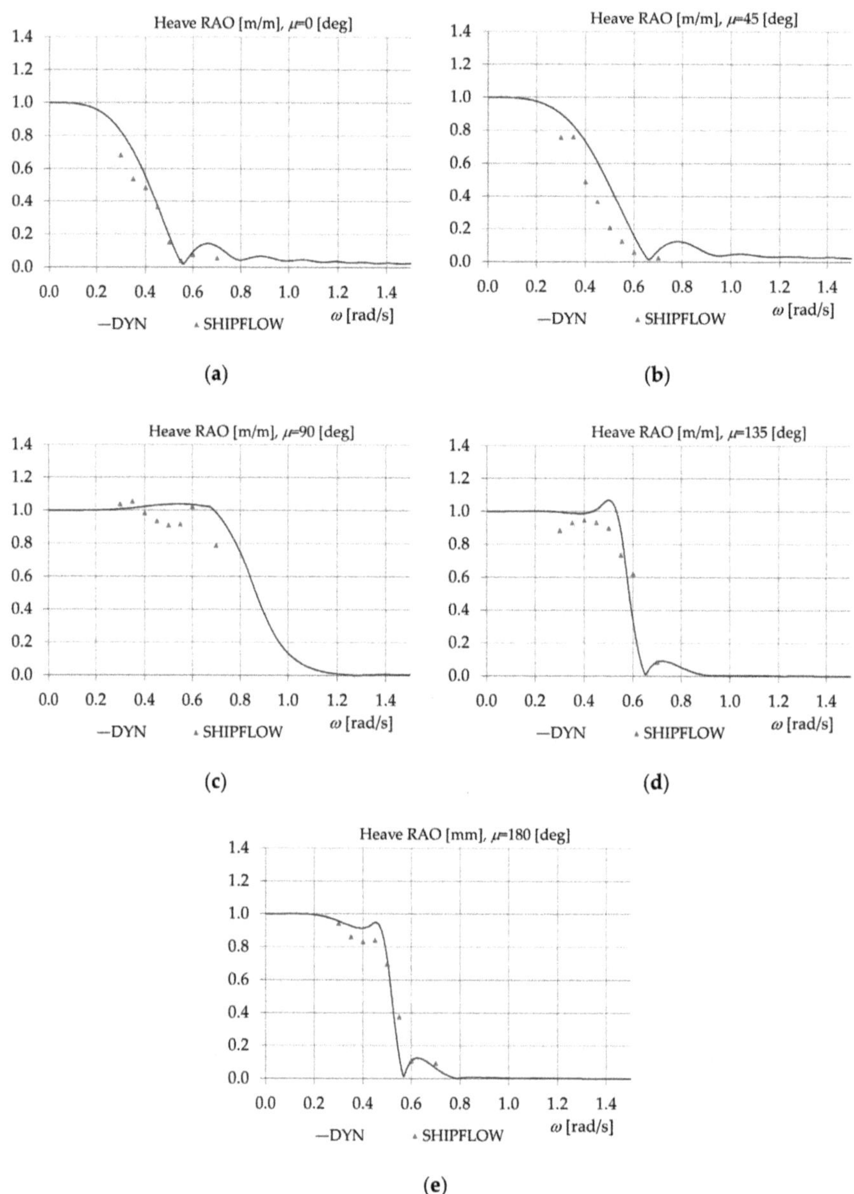

Figure 17. Heave RAO (m/m), v = 24 Kn, h_w = 8 m for the BEM model. (**a**) μ = 0 deg; (**b**) μ = 45 deg; (**c**) μ = 90 deg; (**d**) μ = 135 deg; (**e**) μ = 180 deg.

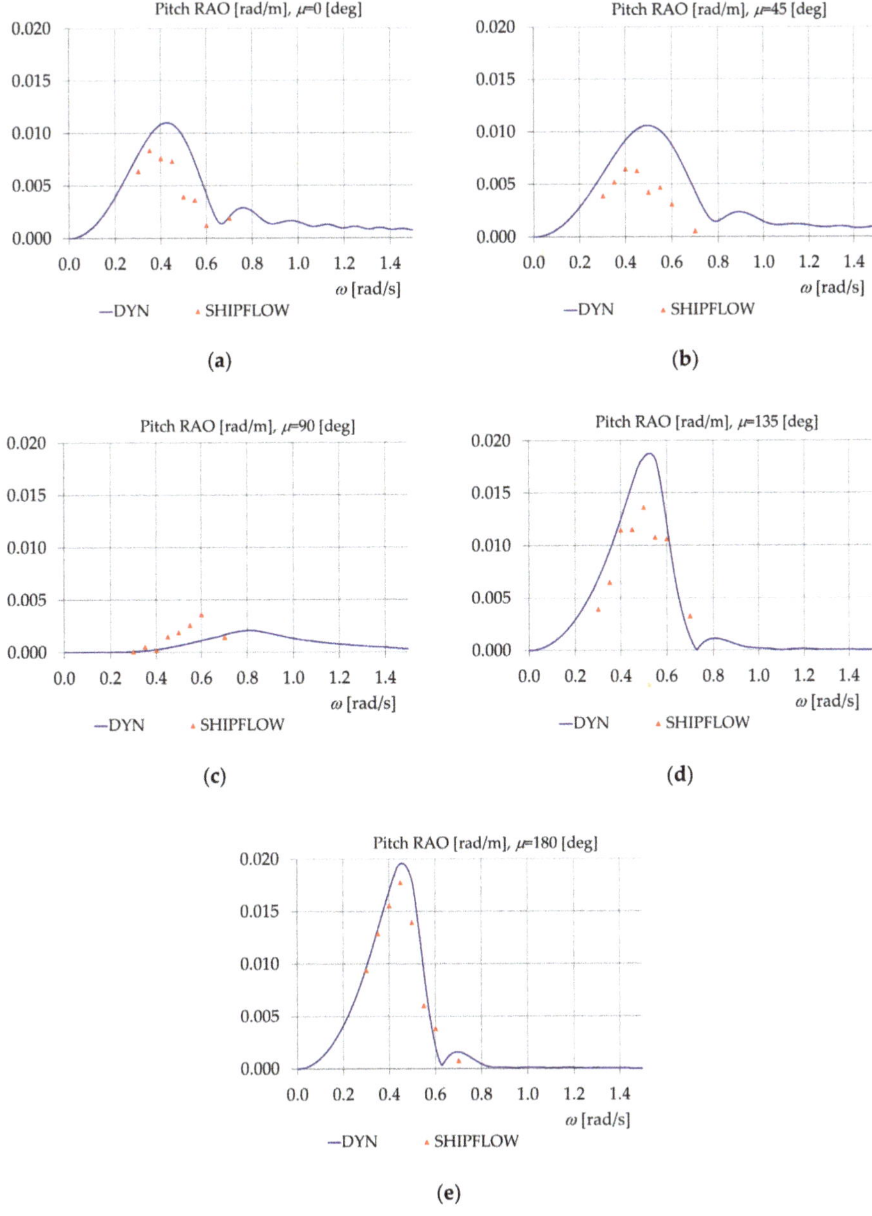

Figure 18. Pitch RAO (rad/m), $v = 24$ Kn, $h_w = 8$ m for the BEM model. (**a**) $\mu = 0$ deg; (**b**) $\mu = 45$ deg; (**c**) $\mu = 90$ deg; (**d**) $\mu = 135$ deg; (**e**) $\mu = 180$ deg.

The Heave RAO peak is 1.071 m/m around $\omega = 0.5$ rad/s for $\mu = 135$ deg (as can be seen in Figure 17) while the maximum value of the Pitch RAO is 0.02 rad/m around $\omega = 0.45$ rad/s (shown in Figure 18), when the ship is in head waves $\mu = 180$ deg.

In Figure 19, a comparison between heave amplitude spectrum, by DFT (Direct Fourier Transformation) method implemented in DYN code, for the two speed cases (12 and 24 Kn) and wave height h_w = 2 m, heading angle μ = 180 deg, wave circular frequency ω = 0.4 rad/s (with ship–wave f_e encountering frequencies 0.080 and 0.096 Hz), is made. The same comparison for pitch amplitude spectrum, by DFT method is made. Both diagrams represent the results obtained using the BEM boundary element method by SHIPFLOW. In Figure 20, the same comparison is made for wave height of h_w = 8 m.

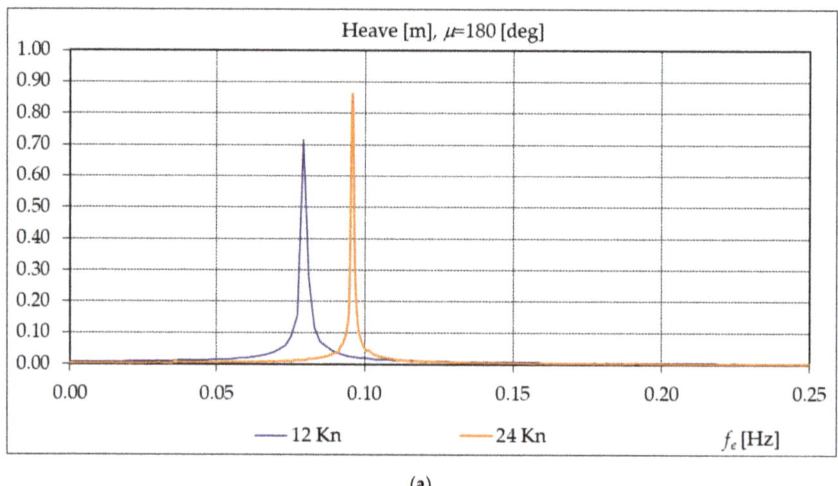

(a)

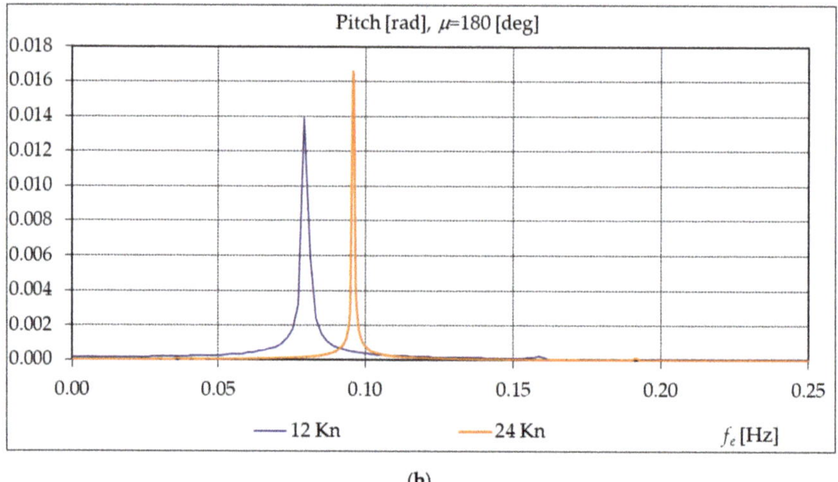

(b)

Figure 19. SHIPFLOW—Amplitude Spectrum by DFT method, v = 12 and 24 Kn, h_w = 2 m, μ = 180 deg, ω = 0.4 rad/s (f_e = 0.080 and 0.096 Hz). (**a**) Heave Amplitude Spectrum (m); (**b**) Pitch Amplitude Spectrum (rad).

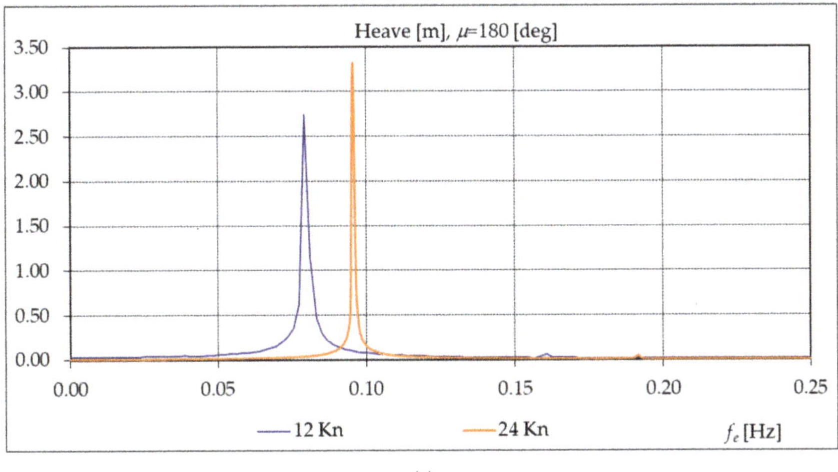

(**a**)

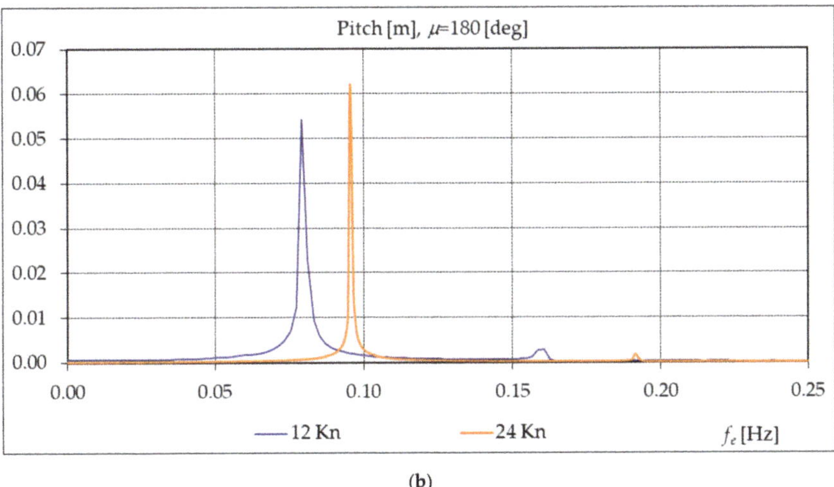

(**b**)

Figure 20. SHIPFLOW— Amplitude Spectrum by DFT method, $v = 12$ and 24 Kn, $h_w = 8$ m, $\mu = 180$ deg, $\omega = 0.4$ rad/s ($f_e = 0.080$ and 0.096 Hz). (**a**) Heave Amplitude Spectrum (m); (**b**) Pitch Amplitude Spectrum (rad).

Both heave and pitch amplitude spectra by DFT method, in the case of higher ship speed of 24 Kn, have narrowed frequency bands. Additionally, at $h_w = 8$ m wave height, the heave and pitch amplitude spectra present a series of higher harmonics, with small amplitudes, due to the hydrodynamic nonlinearities.

In Figures 21 and 22, a comparison between BEM and RANS hydrodynamic models is performed. A comparison of Heave Time Record and Pitch Time Record computed for the KCS vessel at $v = 12$ Kn at heading angle $\mu = 180$ deg and circular wave frequency $\omega = 0.4$ rad/s ($f_e = 0.080$ Hz) is presented in Figure 21.

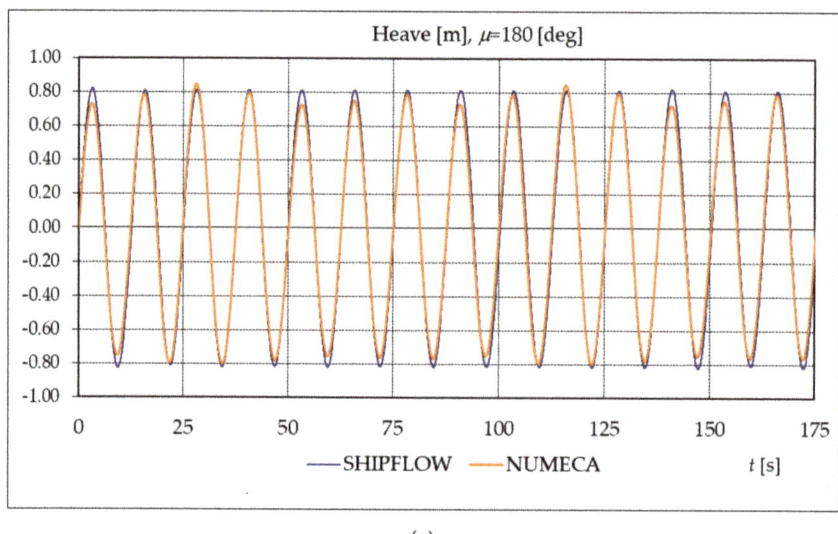

(a)

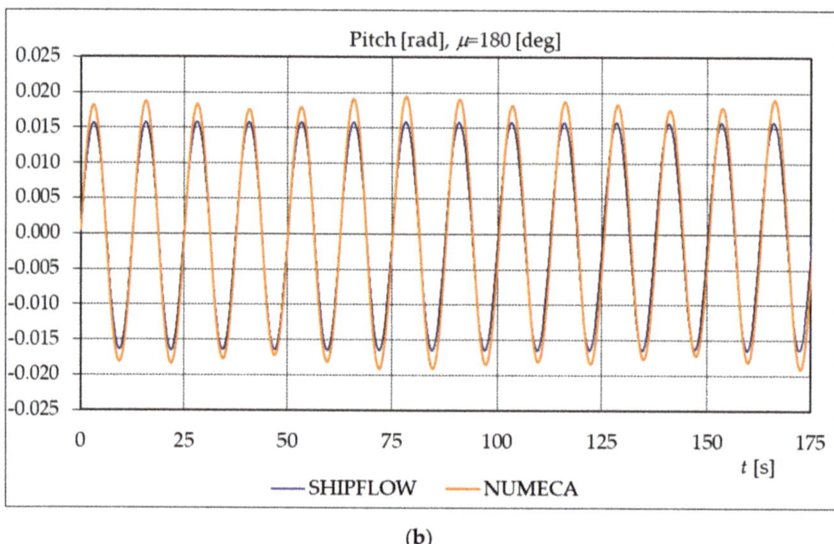

(b)

Figure 21. NUMECA–SHIPFLOW Time Records comparison v = 12 Kn, h_w = 2 m, μ = 180 deg, ω = 0.4 rad/s (f_e = 0.080). (**a**) Heave Time Record (m) (**b**) Pitch Time Record (rad).

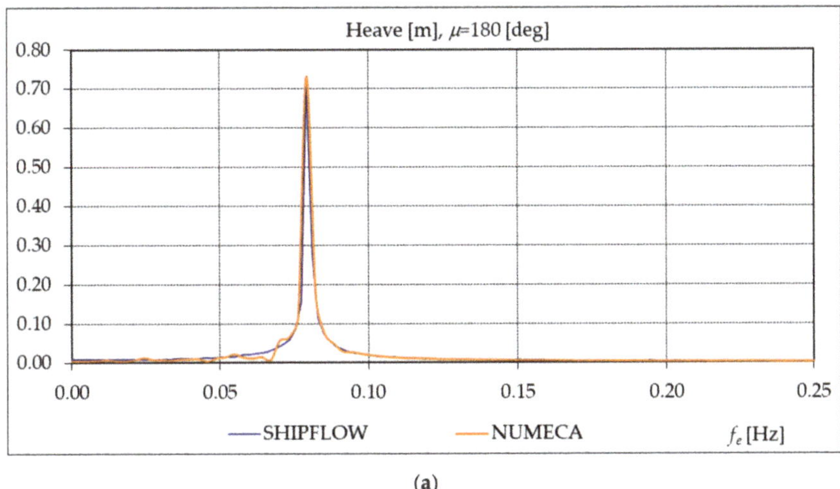

(**a**)

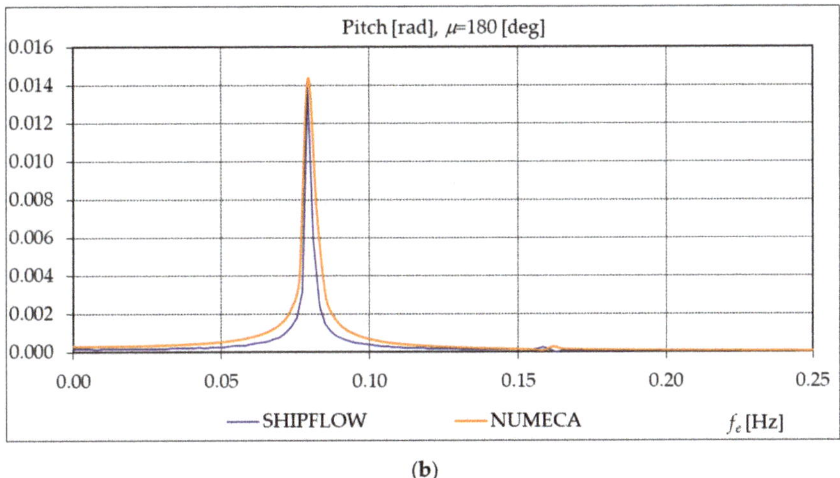

(**b**)

Figure 22. Comparison of NUMECA–SHIPFLOW Amplitude Spectrum by DFT method, $v = 12$ Kn, $h_w = 2$ m, $\mu = 180$ deg, $\omega = 0.4$ rad/s ($f_e = 0.080$). (**a**) Heave Amplitude Spectrum [m]; (**b**) Pitch Amplitude Spectrum (rad).

A comparison of Heave Amplitude Spectrum and Pitch Amplitude Spectrum computed for the KCS vessel at $v = 12$ Kn, at heading angle $\mu = 180$ deg and circular wave frequency $\omega = 0.4$ rad/s ($f_e = 0.080$ Hz) is presented in Figure 22. While for the Heave Amplitude Spectrum, SHIPFLOW and NUMECA show good agreement, for the Pitch Amplitude Spectrum some difference occurs between the two models.

7. Conclusions

In this paper, for the KCS full-scale model (Section 3), several numerical seakeeping analyses have been performed, in order to compare three hydrodynamic methods for the prediction of the dynamic response on heave, pitch and roll oscillations, in terms of response amplitude operators. The selected hydrodynamic methods are the most used approaches for ships motions, linear 2D strip theory, BEM and RANS methods, involving in-house codes and commercial codes.

A grid convergence study has been conducted to determine the accuracy of the grid used to compute the flow based on the potential BEM approach (Section 5). The study revealed that solution computed by SHIPFLOW is grid independent and the grid error for the grid used is less than 1%.

The BEM and RANS methods, with time-domain solutions, were compared to the linear strip (ST) method, with frequency-domain solution; although the external excitation is a regular wave, the ship's response shows higher harmonics (Figures 19 and 20), due to the hydrodynamic nonlinearities. Therefore, in order to ensure the same reference for all three methods, the nonlinear response amplitude operators (RAOs), obtained using BEM and RANS methods, were computed considering only the main spectral amplitude for the ship-wave encountering frequency, being the only result delivered by the linear 2D strip method. Nevertheless, the BEM and RANS studies have revealed, for the KCS model, reduced nonlinear dynamic response components, even if the regular wave height was up to 8 m and the speed of the ship was 24 Kn (Figures 19, 20 and 22).

For a detailed benchmark between the three hydrodynamic models (Section 2), this parametric study developed for the KCS model (Sections 3 and 4), at one loading case, has been focused on the influence of ships speed (12 and 24 Kn), wave heading angle (0, 45, 90, 135, 180 deg) and height (2 and 8 m), on the ship's dynamic response in regular wave, using deterministic analyses (Section 6).

Comparing the Heave RAO functions between the three hydrodynamic models' results (Figures 7, 9, 12, 14 and 17) (Section 6), for ship speed of 24 Kn versus 12 Kn and 8 m wave height versus 2 m, the BEM method delivers visible differences due to the hydrodynamic nonlinearities, especially for the wave circular frequencies up to 0.6 rad/s. In the case of beam waves, for wave circular frequency within the range of 0.4 to 0.6 rad/s, the BEM values are quite scattered around the 2D strip model results. In addition, visible differences, for head waves, have been seen between the RANS method and the other two (Figure 9).

Comparing the Pitch RAO functions between the three hydrodynamic models (Figures 8, 10, 13, 15 and 18) (Section 6), it turned out that the maximum values are obtained for head wave conditions and are very small for beam wave conditions. As the ship's speed and wave height increase, the hydrodynamic nonlinearities induce visible differences between the results, especially close to the peak values of Pitch RAO. Except for 0.3 rad/s, on head waves, the differences between the RANS and the other two methods are quite scattered for the analyzed wave's frequency domain (Figure 10).

Comparing the Roll RAO functions between the BEM and 2D strip hydrodynamic results (Figures 11 and 16) (Section 6), it turned out that the maximum values are obtained for beam waves condition, being a narrow frequency band response amplitude operator. The differences are noticeable at a beam wave condition close to the RAO peak value, especially due to the different hydrodynamic damping formulation of the two methods.

As an overall analysis, we can conclude that the RAO's values predicted by the linear 2D strip theory can be considered as average values compared to the BEM and RANS results, being suitable for practical seakeeping analyses of mono-hull slender body ships.

Further studies will extend this parametric analysis to more numerical results based on the RANS method considering other ship types, being focused on the sensitivity of the hydrodynamic methods used for ship's motions prediction. Additionally, the next step consists of short-term seakeeping analysis in irregular waves, where, besides the BEM and RANS methods, also a nonlinear strip theory with time-domain solution is to be applied.

Author Contributions: Conceptualization, F.P. and L.D.; methodology, F.P., L.D. and S.P.; software, F.P., L.D. and S.P.; validation, F.P. and L.D.; formal analysis, S.P.; investigation, F.P., L.D. and S.P.; resources, F.P., L.D. and S.P.; data curation, F.P.; writing—original draft preparation, F.P., L.D. and S.P.; writing—review and editing, S.P., F.P. and L.D.; visualization, S.P. and F.P.; supervision, L.D.; project administration, S.P.; funding acquisition, S.P. All authors have read and agreed to the published version of the manuscript.

Funding: This research received no external funding.

Acknowledgments: The research study presented in this article was developed in the frame of Dunarea de Jos University of Galati, Naval Architecture Research Centre.

Conflicts of Interest: The author declares no conflict of interest.

Abbreviations

KCS	KRISO container ship
KRISO	Korea Research Institute of Ships and Ocean Engineering
IMO	International Maritime Organization
DYN	Dynamic Ship Analysis
RANSE	Reynolds-averaged Navier–Stokes equation
BEM	Boundary element method
ST	Strip theory
EEOI	Energy efficiency operational indicator
CFD	Computational fluid dynamics
RAO	Response amplitude operator
KVLCC	KRISO Very Large Crude Carrier
OSC	Oscillations
DOF	Degrees of freedom
DFT	Direct Fourier Transformation
Nomenclature	
k-ω SST	k-ω turbulence model, SST shear stress transport
$Lref$	Reference length
L_{BP}	Length between the perpendiculars
L_{WL}	Length of waterline
B_{WL}	Beam at waterline
D	Depth
T	Design draft
Δ	Displacement
C_B	Block coefficient
v	Ship's speed
L_{CB}	Longitudinal center of buoyancy ($\%L_{BP}$), fwd+
L_{CG}	Longitudinal center of gravity from the aft peak
KG	Vertical center of gravity from keel
Kxx/B	Moment of inertia
Kyy/L_{BP}, Kzz/L_{BP}	Moment of inertia
S	Ship wetted area
ω	Wave circular frequency
f_e	Ship-wave encountering frequency
h_w	Wave height
a_w	Wave amplitude
λ	Wavelength
μ	Heading angle
ε	Simulation error
R_G	Convergence ratio
Si	Solution calculated for fine (S_1), medium (S_2) and coarse grid (S_3)
r_G	Refinement ration
p_G	Order of accuracy
C_G	Correction factor
δG	Error
p_{th}	Theoretical grid convergence order

References

1. Shukui, L.; Papanikolau, A. On Nonlinear Simulation Methods and Tools for Evaluating the performance of Ships and Offshore Structures in Waves. *J. Appl. Math.* **2012**, *2012*, 1–21.
2. Beck, R. Modern Seakeeping Computations for Ships. In *Twenty-Third Symposium on Naval Hydrodynamics*; The National Academies Press: Washington, DC, USA, 2001. [CrossRef]
3. ITTC, The Seakeeping Committee. Final report and recommendations to the 25th ITTC. In Proceedings of the 25th ITTC, Fukuoka, Japan, 14–20 September 2008; Volume I.
4. Bertram, V. *Practical Ship Hydrodynamics*, 2nd ed.; Butterworth-Heinemann: Oxford, UK, 2012.
5. Sclavounos, P.D.; Lee, C. Topics on Boundary Element Solutions of Wave Radiation-Diffraction Problems. In Proceedings of the 4th International Conference on Numerical Ship Hydrodynamics, Washington, DC, USA, 24–27 September 1988.
6. Korvin-Kroukovsky, B. Investigation of ship motions in regular waves. *Trans. SNAME* **1955**, *63*, 386–435.
7. Salvesen, N.; Tuck, E.O.; Faltinsen, O.M. Ship motions and sealoads. *Trans. SNAME* **1970**, *78*, 250–287.
8. Gerritsma, J.; Beukelman, W. Analysis of the Modified Strip Theory for Calculation of Ship Motions and Wave Bending Moments. *Int. Shipbuilding Prog.* **1967**, *14*, 319–337. [CrossRef]
9. Ogilvie, T.F.; Tuck, E.O. *Rational Strip Theory of Ship Motions*; University of Michigan: Ann Arbor, MI, USA, 1969.
10. Holloway, D.; Davis, M. Ship Motion Computations Using a High Froude Number Time Domain Strip Theory. *J. Ship Res.* **2006**, *50*, 15–30.
11. Jensen, J.J.; Pedersen, P.T.; Shi, B.; Wang, S.; Petricic, M.; Mansour, A.E. *Wave Induced Extreme Hull Girder Loads on Containership*; The Society of Naval Architects and Marine Engineers Transactions: Jersey City, NJ, USA, 2008; Volume 116.
12. Sutulo, S.; Guedes Soares, C. A Generalized Strip Theory for Curvilinear Motion in Waves. In Proceedings of the 27th International Conference on Offshore Mechanics and Arctic Engineering, Estoril, Portugal, 15–20 June 2008; pp. 359–368.
13. Prpic-Orsic, J.; Faltinsen, O.M. Estimation of ship speed loss and associated CO_2 emissions in a seaway. *Ocean Eng.* **2012**, *44*, 1–10. [CrossRef]
14. Bandyk, P.J.; Hazen, G.S. A forward-speed body-exact strip theory. In Proceedings of the 34th International Conference on Ocean, Offshore and Arctic Engineering, St. John's, NL, Canada, 31 May–5 June 2015.
15. Newman, J.N. *Marine Hydrodynamics*, 1st ed.; MIT Press: Cambridge, MA, USA, 1977.
16. Faltinsen, O.M. *Hydrodynamics of High-Speed Marine Vehicles*, 1st ed.; Cambridge University Press: Cambridge, UK, 2005.
17. Lin, H.; Lin, C.W. Numerical Simulation of Seakeeping Performance on the Preliminary Design of a Semi-Planing Craft. *J. Mar. Sci. Eng.* **2019**, *7*, 199. [CrossRef]
18. Rorvik, J. Application of Inviscid Flow CFD for Prediction of Motions and Added Resistance of Ships. Master's Thesis, NTNU Trondheim, Trondheim, Norway, 2016.
19. Hess, J.L.; Smith, M.O. Calculation of Nonlifting Potential Flow about Arbitrary Three-dimensional body. *J. Ship Res.* **1964**, *8*, 22–44.
20. Gadd, G.E. *A Method of Computing the Flow and Surface Wave Pattern around Full Forms*; The Royal Institution of Naval Architects: London, UK, 1976; pp. 207–219.
21. Dawson, C.W. A Practical Computer Method for Solving Ship-Wave Problem. In Proceedings of the 2nd International Conference on Numerical Ship Hydrodynamics, Berkeley, CA, USA, 19–21 September 1977; pp. 30–38.
22. Jensen, G.; Mi, Z.X.; Söding, H. Rankine Source Methods for Numerical Solutions of the Steady Wave Resistance Problem. In Proceedings of the 16th Symposium on Naval Hydrodynamics, Berkeley, CA, USA, 13–18 July 1986.
23. Raven, H.C. Variations on a Theme by Dawson. In Proceedings of the 17th Symposium on Naval Hydrodynamics, The Hague, The Netherlands, 29 August–2 September 1988; pp. 151–172.
24. Sclavounos, P.D.; Nakos, D.E. Stability Analysis of Panel Methods for Free Surface Flow with forward Speed. In Proceedings of the 17th Symposium on Naval Hydrodynamics, The Hague, The Netherlands, 29 August–2 September 1988.

25. Janson, C.E. Potential Flow Panel Methods for the Calculation of Free-Surface Flows with Lift. Ph.D. Thesis, Chalmers University of Technology, Gothenburg, Sweden, 1997.
26. Nakos, D.E.; Kring, D.E.; Sclavounos, P.D. Rankine Panel Method for Time Domain Free Surface Flows. In Proceedings of the 6th International Conference on Numerical Ship Hydrodynamics, Iowa City, IA, USA, 2–5 August 1993.
27. Zhang, X.; Bandyk, P.; Beck, R.F. Large Amplitude Body Motion Computations in the Time-Domain. In Proceedings of the 9th International Conference in Numerical Ship Hydrodynamics, Ann Arbor, MI, USA, 5–8 August 2007.
28. Söding, H.; von Gräfe, A.; el Moctar, O.; Shigunov, V. Rankine Source Method for Seakeeping Predictions. In Proceedings of the 31st International Conference on Ocean, Offshore and Arctic Engineering, Rio de Janeiro, Brazil, 1–6 July 2012.
29. Mei, T.; Candries, M.; Lataire, E.; Zou, Z. Numerical Study on Hydrodynamics of Ships with Forward Speed Based on Nonlinear Steady Wave. *J. Mar. Sci. Eng.* **2020**, *8*, 106. [CrossRef]
30. Dai, Y.Z.; Wu, G.X. Time Domain Computation of Large Amplitude Body Motion with the Mixed Source Formulation. In Proceedings of the 8th ICHD, Nantes, France, 30 September–3 October 2008; pp. 441–452.
31. Deng, D.G.; Queutey, P.; Visonneau, M. RANS prediction of KVLCC2 tanker in head waves. In Proceedings of the 9th International Conference on Hydrodynamics, Shanghai, China, 11–15 October 2010.
32. Choi, J.; Yoon, S.B. Numerical simulations using momentum source wave-maker applied to RANS equation model. *Coast. Eng.* **2009**, *56*, 1043–1060. [CrossRef]
33. Sato, Y.; Miyata, H.; Sato, T.J. CFD simulation of 3-dimensional motion of a ship in waves: Application to an advancing ship in regular heading waves. *J. Mar. Sci. Technol.* **1999**, *4*, 108–116. [CrossRef]
34. Simonsen, C.D.; Otzen, J.F.; Stern, F. EFD and CFD for KCS Heaving and Pitching in Regular Head Waves. *J. Mar. Sc. Tech.* **2013**, *18*, 435–459. [CrossRef]
35. Lungu, A. Unsteady Numerical Simulation of the Behaviour of a Ship Moving in Head Sea. In Proceedings of the 38th International Conference on Ocean, Offshore and Arctic Engineering, Glasgow, UK, 9–14 June 2019.
36. Beckhit, A.; Lungu, A. URANSE simulation for the Seakeeping of the KVLCC2 Ship Model in Short and Long Regular Head Waves. *IOP Conf. Ser. Mater. Sci. Eng.* **2019**, *591*, 012102. [CrossRef]
37. Hochkirch, K.; Mallol, B. On the importance of full-scale CFD simulations for ships. In Proceedings of the 12th International Conference on Computer Applications and Information Technology in the Maritime Industries (COMPIT 2013), Cortona, Italy, 15–17 April 2013; pp. 85–95.
38. Tezdogan, T.; Demirel, Y.K.; Kellett, P.; Khorasanchi, M.; Incecik, A.; Turan, O. Full-scale unsteady RANS CFD simulations of ship behaviour and performance in head seas due to slow steaming. *Ocean Eng.* **2015**, *97*, 186–206. [CrossRef]
39. Domnisoru, L. *Ship Dynamics. Oscillations and Vibrations*; Technical Publishing House: Bucharest, Romania, 2001.
40. Kjellberg, M.; Janson, C.E.; Contento, G. Fully Nonlinear Potential Flow Method for Three-Dimensional Body Motion Proceedings from NAV 2012. In Proceedings of the 17th International Conference on Ships and Shipping Research, Naples, Italy, 17–19 October 2012.
41. Duvigneau, R.; Visonneau, M.; Deng, G.B. On the role played by turbulence closures in hull shape optimization at model and full-scale. *J. Mar. Sci. Technol.* **2003**, *8*, 1–25. [CrossRef]
42. Queutey, P.; Visonneau, M. An interface capturing method for free-surface hydrodynamic flows. *Comput. Fluids* **2007**, *36*, 1481–1510. [CrossRef]
43. Obreja, D.; Nabergoj, R.; Crudu, L.; Domnisoru, L. Seakeeping performance of a Mediterranean fishing vessel. In Proceedings of the IMAM Maritime Transportation and Harvesting of Sea, Lisbon, Portugal, 9–11 October 2017; pp. 483–491.
44. Burlacu, E.; Domnisoru, L.; Obreja, D. Seakeeping Prediction of a Survey Vessel Operating in the Caspian Sea. In Proceedings of the 37th International Conference on Offshore Mechanics and Arctic Engineering, Madrid, Spain, 17–22 June 2018.
45. Zhang, B.J.; Ning, X. The research of added resistance in waves based on nonlinear time-domain potential flow theory. *J. Mar. Sci. Technol.* **2018**, *26*, 343–351.
46. Larsson, L.; Stern, F.; Visonneau, M. *Numerical Ship Hydrodynamics—An Assessment of the Gothenburg 2010 Workshop*; Springer: Dordrecht, The Netherlands, 2013.

J. Mar. Sci. Eng. **2020**, *8*, 962

47. Coslovich, F.; Contento, G.; Kjellberg, M.; Janson, C.E. *Computations of Roll Motions in Waves Using a Fully Nonlinear Potential Flow Method, Technologyand Science for the Ship of the Future*; Marino, A., Bucci, V., Eds.; IOS press: Amsterdam, The Netherlands, 2018; pp. 186–193.
48. Lungu, A. Numerical simulation of the resistance and self-propulsion model tests. *J. Offshore Mech. Arctic Eng.* **2020**, *142*, 021905. [CrossRef]
49. Bekhit, A.; Lungu, A. Numerical Simulation for Predicting Ship Resistance and Vertical Motions in Regular Head Waves, ASME 2019. In Proceedings of the 38th International Conference on Ocean, Offshore and Arctic Engineering, Glasgow, Scotland, 9–14 June 2019.
50. Pacuraru, S.; Domnisoru, L.; Pacuraru, F. Numerical study on motions of a containership on head waves. In Proceedings of the 6th International Conference—Modern Technologies in Industrial Engineering, Constanta, Romania, 13–16 June 2018; Volume 400.
51. Kjellberg, M. Fully Nonlinear Unsteady Three-Dimensional Boundary Element Method for Ship Motions in Waves. Ph.D. Thesis, Chalmers University of Technology, Gothenburg, Sweden, 2013.
52. Coslovich, F. Computations of Ship Motions in Waves Using a Fully Nonlinear Time Domain Potential Flow Method. Master's Thesis, University of Trieste, Trieste, Italy, 2015.
53. International Towing Tank Conference (ITTC), Uncertainty Analysis in CFD Verification and Validation Methodology and Procedures 7.5-03-01-01. 2017, pp. 1–13. Available online: https://www.ittc.info/media/8153/75-03-01-01.pdf (accessed on 20 August 2020).

Publisher's Note: MDPI stays neutral with regard to jurisdictional claims in published maps and institutional affiliations.

MDPI

St. Alban-Anlage 66

4052 Basel

Switzerland

Tel. +41 61 683 77 34

Fax +41 61 302 89 18

www.mdpi.com

Journal of Marine Science and Engineering Editorial Office

E-mail: jmse@mdpi.com

www.mdpi.com/journal/jmse